URBAN POLICY ANALYSIS

Directions for Future Research

Volume 21, URBAN AFFAIRS ANNUAL REVIEWS

URBAN POLICY ANALYSIS

Directions for Future Research

Edited by
TERRY NICHOLS CLARK

Volume 21, URBAN AFFAIRS ANNUAL REVIEWS

SAGE PUBLICATIONS Beverly Hills London

For information address:

SAGE Publications, Inc.
275 South Beverly Drive
Beverly Hills, California 90212

SAGE Publications Ltd
28 Banner Street
London EC1Y 8QE, England

Printed in the United States of America

Library of Congress Cataloging in Publication Data

Main entry under title:
Urban policy analysis.

(Urban affairs annual reviews ; v. 21)
Includes edited papers presented at the Conference on Comparative Urban Policy Research, held Apr. 26-27, 1981, University of Chicago, which was supported by the National Science Foundation.
Bibliography: p.
1. Urban policy—Research—United States—Congresses.
I. Clark, Terry N. II. Conference on Comparative Urban Policy Research (1981 : University of Chicago) III. Series.
Ht108.U7 vol. 21 [HT123] 307.7'6s 81-14404
ISBN 0-8039-1627-2 [307.7'6072073] AACR2
ISBN 0-8039-1628-0 (pbk.)

FIRST PRINTING

Contents

Preface

A CONFERENCE helped initiate this volume. Twenty persons presented papers and discussed these issues at the Conference on Comparative Urban Policy Research held April 26-27, 1981 on the University of Chicago campus, supported by the National Science Foundation.[1] Conference participants and papers were as follows:

Robert L. Lineberry, Northwestern University, "New Wine in Old Paradigms"

Paul D. Schumaker, University of Kansas, "Theoretical and Methodological Issues Regarding the Investigation of Citizen Preferences and Policy Responsiveness Through Comparative Urban Research."

John R. Logan, SUNY Stony Brook, "Metropolitan Political Economy"

Mark Schneider, SUNY Stony Brook, "Population Change and the Fiscal Condition of Metropolitan Communities"

John D. Kasarda, University of North Carolina at Chapel Hill, "Population and Economic Base Changes in Metropolitan Areas"

Peter H. Rossi, University of Massachusetts, "Estimating Program Impacts Using Time Linked Small Area Census Data"

Terry Nichols Clark and Lorna Crowley Ferguson, National Opinion Research Center and University of Chicago, "Political Leadership and Urban Fiscal Policy"

David Knoke, Indiana University, "Urban Political Cultures"

Roger Friedland and William T. Bielby, University of California, Santa Barbara, "The Power of Business in the City"

Ronald S. Burt, University of California, Berkeley and SUNY Albany, "Spatial Models of Community Leadership," and "Comparative Power Structures in American Communities"

Elinor Ostrom, Indiana University, "Productivity in the Urban Public Sector," and "Urban Resources, Institutions, and Outcomes"

Thomas J. Anton, University of Michigan, "Data Systems for Urban Fiscal Policy"

George S. Tolley and Larry E. Huckins, University of Chicago, "Capital Requirements and the Future of Cities" and "Investments in Local Public Infrastructure"

Other participants included:

Roland J. Liebert, Program Director for Sociology, National Science Foundation

Kenneth Prewitt, Director, National Opinion Research Center, University of Chicago, and President, Social Science Research Council

Gerald D. Wright, Program Director for Political Science, National Science Foundation

Thomas Smith, National Opinion Research Center

Michael Traugott, Inter-University Consortium for Social and Political Research

Each paper reviewed a subarea of urban research, stated some principal theoretical questions in the subarea, and indicated specific types of data that could be used to address these questions. Following discussion at the conference, the papers were revised or rewritten to incorporate suggestions and criticisms; the revised versions appear in Parts II through V of this volume. Important contributions in their own right, the papers take on further interest due to their preparation within a common framework.

The effort to identify common elements began at the conference and continued initially in a Conference Summary (June 1979) that was circulated to numerous urban researchers. The themes were then discussed by several of us with urban researchers at professional meetings. The next step was for seven of us to write a statement of common interest that would sharpen the focus by identifying common themes, research hypotheses, and critical data to collect. The result, "Urban Policy Analysis: A New Research Agenda," constitutes Chapter 2 of this volume. It was submitted to the National Science Foundation as a proposal for a continuing research project. While some reviewers showed considerable enthusiasm (and several suggested the proposal be published), the magnitude of the effort was such that it was deferred. As this is being written, federal funding for urban research is very much in flux. To undertake a project of the kind outlined here is obviously more ambitious than many; but so are its payoffs. Just how the themes considered might best be adapted to changing concerns of funding agencies and policy-oriented urban researchers clearly demands discussion by interested persons. This volume is presented to help stimulate such discussion.

Chicago, Illinois — *Terry Nichols Clark*

NOTE

1. "Comparative" has not been retained in the title for this volume, despite the international audience of Sage *Urban Affairs Annual Reviews.* In fact, although we generally focus on American cities, we have sought to unravel important questions by comparing them with each other. "Policy" generally refers to decisions made by public officials; and "policy-oriented" research or analysis, to studies that might inform such decisions. The scope of the volume is considered further in Chapter 1.

The Center for Urban Studies of the University of Chicago helped arrange for the conference, and Kenneth Prewitt offered a reception.

Part I

Introduction

1

An Overview of the Volume

TERRY NICHOLS CLARK

IN THE LAST DECADE, elected officials in many cities, state capitals, and Washington have grown far more sophisticated in their concerns for useful information about the urban programs that they help design and fund. Simultaneously, researchers have developed more powerful techniques for addressing critical policy issues. Elected and administrative officials were less likely to request sophisticated analyses just a decade or so ago, and researchers were less equipped to respond to their concerns. Sufficient work has now been completed using data from ad hoc studies, the census, and other varied sources that the potential for more significant contributions has become clear. Here are some examples of important research questions elaborated below:

• What are the sources and solutions of the taxpayers' revolt? Which cities have been able to adapt policies to the concerns of their constituents? What factors lead cities toward or away from such policies? How much turnover in elected officials is there in different cities and over time? What are some of the causes of high turnover? How have citizens' and leaders' policy preferences shifted in relation to each other over time?

• How fiscally strained are different American cities? Why have some older, Northeastern cities that one might expect to be fiscally strained (such as Pittsburgh) been able to adapt and reduce fiscal strain, while others (such as New York) have had more difficulties?

• How can public service agencies be organized to deliver better services at lower costs? What are the effects on services of different types of subcontracting, of intergovernmental coordination, of centralization of authority, of salary and promotion criteria, of relaxing civil service regulations?

• What are the specific interrelations between movement of jobs and population? How much effort should cities put into attracting residents as opposed to firms? How about large versus small firms? How much of an impact do tax burdens and service delivery have on location decisions by residents and firms?

These are critical questions for federal and state officials concerned with cities, for city officials, and for urban researchers. Answers from the best research to date remain divided, but billions of dollars might be reallocated if answers were clearer.

Many contributors to the volume have been active in both basic research and more policy-oriented work. The distinction between the two types of research is sometimes significant; other times it is not. The issues considered in the volume are often central to policy-oriented questions as well as to more basic research.

ORGANIZATION OF THIS VOLUME

These major themes are reviewed in Chapter 2 ("Urban Policy Analysis: A New Research Agenda") and subsequent chapters. The Agenda (in an early section, headed "Four Foci for Research") indicates linkages among the four basic processes considered in the volume. The four are sufficiently interrelated to constitute a coherent research agenda, but broad enough to capture the major factors generating and influenced by urban public policies. While several papers elaborate propositions moving toward an integrated theory, our concern to keep the enterprise open to alternative perspectives led us to demote theoretical coherence to a secondary goal. Nevertheless, these four specific areas of emphasis were chosen because they had developed more coherent bodies of knowledge than most in the social sciences. Our goal is to stimulate future research by outlining several dozen suggestive hypotheses, but to do so in such a manner that researchers of many different perspectives can be challenged to build on them.

The four next sections of this volume summarize, first, the corresponding portion of the Agenda paper, and then individual papers in that section.

URBAN POLITICS AND POLICY OUTPUTS

The Agenda outlines several theories that link political leadership processes to policy outputs of the political system. *Elite theories* stress social background characteristics of leaders. *Network analysis* considers linkages among social, economic, and political leaders. *Organized group analysis* focuses on resources and influence processes of specific groups. *Coalition theory* considers how and why different groups and leaders join to affect public policy. *Citizen and leader preferences* concerning specific policies are the focus of spatial modeling work, while *political culture* usually concerns more general rules of game. *Bureaucratic theories* posit incrementalist rules or self-aggrandizing bureau heads as direct sources of policy. Surveying a national sample of urban leaders every two years or so would generate a rich store of data by which such alternative theories could be assessed; given current survey techniques, the cost would be reasonable for perhaps a dozen leaders in each of 50 to 100 cities. The Agenda lists a series of variables that would permit analysis of the alternative theories.

Individual chapters in this section consider related themes in more detail. Clark and Ferguson outline a theory of contextual relativism focusing on variables identified by the theories listed above. Rather than suggesting that the theories are right or wrong, they suggest when and where the variables they identify are differentially important in affecting policy outputs. Uncertainty in the minds of political leaders concerning citizen preferences is a key shifter variable, in that when leaders have complete imformation, one would expect them to become "invisible politicians," implementing policies preferred by citizens just as the invisible hand responds to consumer preferences in the private market. But seldom is information complete; as uncertainty concerning citizen preferences increases, leaders are more likely to act on their own personal preferences or to respond to bureau heads or leaders of organized groups in formulating policies. The political culture of a city acts as a resource facilitating certain types of coalitions and the ability of some actors to achieve their goals while discouraging others. Clark and Ferguson then outline a series of specific options concerning data collection procedures that, when applied to a national sample of cities, would permit testing theories of the kind proposed. They illustrate concretely how competing theories and findings might be reconciled through analysis of comparable data for a large sample of cities.

Burt focuses on two bodies of theory that have developed significantly in the last decade; spatial modeling and network analysis. Most work in each tradition has ignored the other. But the two offer complementary insights into policy formation; joining them can advance each tradition while generating a more balanced empirical interpretation. Spatial modeling stresses specific policy preferences of citizens and leaders and how they encourage coalitions among groups sharing relatively similar policy preferences (as in minimal range theory or policy distance theory). Leaders may variously respond to citizens, as indicated by a loss function of the discrepancy between a leader's policy preferences and those of a constituent. The driving force in such theories (variously labeled "public choice," "spatial modeling," "citizen preference," or "economic theories of democracy") is citizen preference. Political leadership is seen to function essentially as a clean market mechanism matching citizen and leader preferences in policy outputs. But little is indicated about how and why such a market might function if transaction costs increase. Network analyses can help here by focusing on the degree of interconnectedness among leaders, organized groups, and individual citizens. Considerable effort has gone into developing methods for measuring networks: their patterning, density, and overlap. But network analyses often remain relatively contentless; they do not offer as specific indications about which policies should be preferred as does the spatial modeling tradition. And the spatial modeling tradition has remained generally more mathematical and deductive, while network analyses have been more statistical and inductive. The two are sufficiently close and complementary that they can reinforce each other's insights. Burt suggests several examples of propositions that join the two.

Huckins and Tolley point to a gap in past research: capital infrastructure decisions of city governments. When and why do cities build roads, sewers, and sidewalks and variously maintain them? This question was raised with special concern in the late 1970s after cities like New York cut back on maintenance as a response to fiscal strain. Considering the 46 largest U.S. cities from 1955 to 1977, Huckins and Tolley analyze their capital and labor investments. Increases in citizens' income generate increased investments but with surprisingly different lag structures; the capital stock adjusts to long-run equilibrium levels in about 15 years, while labor adjusts in only half as long. Similarly, there were clear tradeoffs between capital and labor, estimated by elasticities of substitution of approximately 2. This points to the importance of explicitly modeling capital-labor tradeoffs as alternative means to service provision—a theme elaborated in Chapter 8 by Parks and Ostrom. Most research has simply ignored capital investment and concentrated on annual expenditures or labor.

Huckins and Tolley then present a model for considering effects of maintenance expenditures on the capital stock. The billions of dollars represented by the capital stock of American cities, they suggest, is far more than adequate to cover pension liabilities, debt payments, and other future obligations of city governments. Nevertheless, current accounting procedures concentrate only on short-term flows of funds, such that few cities have even estimates of the value of their capital stocks, much less data sufficiently precise to permit modeling effects of maintenance expenditures. Engineers' recommendations tend to fill the void, but these are often conservative and not presented in such manner as to encourage alternative maintenance levels as explicit policy choices.

Friedland and Bielby address relationships between business and city governments by reviewing three types of analyses. The first considers public involvement of businessmen directly. Although Hunter and Dahl differed in several respects, they agreed on the importance of collecting direct information concerning the importance of business actors in public decisions. These were mainly case studies which were superseded by comparative studies of community decision-making in the late 1960s. Such comparative work by Clark, Aiken, and others in turn led to a focus on urban socioeconomic characteristics linked to business power and decentralization of decision-making; characteristics included economic diversification, absentee ownership, location of national headquarters, and concentration of coordinative roles. Much of this work ignored specific business interests and possible policy impacts, although Crenson was one exception. A third approach analyzes not business participation in public decisions, but socioeconomic base characteristics, such as retail sales or housing value, and relates these to public decisions as the size of urban renewal programs. This approach depends on the assumption that the political power of business may be inferred from the magnitude of the relationship between socioeconomic characteristics and policy outputs. Each approach thus has its weaknesses, but Friedland and Bielby argue that it would be useful to consider more carefully the impacts of actual services delivered on business location. Exit may be as much as or more important than voice as a means by which business power may be exerted. But this implies collecting more detailed service delivery data and considering them more carefully in terms of their responsiveness to business interests than has occurred to date. Friedman and Bielby thus emphasize the importance of joining the traditions of political leadership and service delivery through a different kind of analysis of better data.

BUREAUCRATIC PROCESSES AND SERVICE DELIVERY

Political leaders may approve fiscal policies, but implementation in specific services varies considerably across cities, hence the importance of data concerning bureaucratic processes and service delivery. The Agenda lists major issues in this area: services as affecting location decisions of firms and individuals, relations between service provision and citizen participation, changes in service delivery when cities perform a broader range of services or serve larger constituencies, impacts on service delivery of contracting with different governments or private firms, and consequences of using citizens as coproducers. More internal bureaucratic processes include when and why the alternative views of the bureau chief as self-aggrandizing bureaucrat or invisible politician are more appropriate and what the impacts are of civil service, unionization, and formal organizational arrangements. How do all of these shift with citizen preferences and political culture? While the range of questions is broad, there is reasonable agreement concerning the data necessary to address most questions. This section of the Agenda thus includes a rather detailed set of variables that might be collected.

Brooks and Lineberry focus on police services and ask how they may vary with political leadership. Reported crime rates have been alleged to vary for reasons of political expediency. Case studies of Newark and Philadelphia suggest curious associations between election timing of mayors and crime rates. But these are only two of ten cities studied as part of a larger project, including data over 31 years. Using the 300 city-years, Brooks and Lineberry consider several hypotheses linking political leadership to crime rates. Despite the emphasis on crime as an issue in many campaigns, they find the timing of mayoral elections unrelated to crime rate changes. However, they find that when new mayors were elected to office, crime rates did significantly change. They offer several observations on analyzing over-time data rather then comparing across cities at the same time. Similarly, their analyses of Newark and Philadelphia show how case studies can permit intense study of specific processes, but the contrasting results in the comparative analysis underscore the difficulties of generalizing from individual cases.

Parks and Ostrom provide a challenging example of how to analyze these issues. If some work in this field has been criticized for its lack of theoretical coherence, Parks and Ostrom demonstrate forcefully that this need not be the case. While they elsewhere make clear the importance of considering alternative approaches, in Chapter 8 they develop a systematic approach building on industrial economics to study police services. But while the formal apparatus for production functions is taken from others, Parks and Ostrom move significantly ahead by including several critical variables. They specify a production function for police output, including the number of policemen, automobiles, and so on. Then they include functions where citizens prefer good services at low costs and managers seek to expand staff but also provide efficient services. These combine a production possibility frontier and utility functions where citizen and bureaucratic preferences join more standard variables. Frontier production function estimates are then generated where the dependent variables are average number of cars on the street and clearances by arrest. Given their interest in scale economies and metropolitan area structure, Parks and Ostrom estimate equations separately for departments of different size and for metropolitan areas with differing numbers of police departments. Their results show diseconomies of scale, in that larger departments have proportionately fewer cars on the street. They also find that in metropolitan areas with more numerous police departments, the most efficient departments have higher levels of output than the more efficient departments in metropolitan areas with few producers. They interpret the last finding as resulting from greater information exchange among public officials and citizens in metropolitan areas with more numerous police departments. Increased information enables and forces police chiefs to act in a more efficient and public-serving manner, a finding consistent with the Clark and Ferguson analysis.

CITIZEN PREFERENCES AND PARTICIPATORY ROLES

The extensive work on citizenship led us to discuss it extensively in the Agenda statement. Two general theoretical approaches are distinguished: the democratic elitist view, which suggests that citizens have weak and inconsistent preferences, and their inputs to the political system are largely via a shared culture with leaders; and a populist view, which assigns a more dynamic role to citizen preferences. Empirical research on citizen preferences has often been social-psychological, due partly to difficulties of combining community characteristics with the national sample surveys on which most research about citizens has been based. An alternative research design could help. Comparing citizen preferences across cities differing in issue salience, media coverage, campaign efforts, and political culture could sharpen analysis and resolve conflicting findings. One issue is how structured are citizen preferences. Samuelson and Arrow offered impossibility theorems, and Converse, empirical results implying problematic structure to citizen policy preferences. Others have suggested more structure, using several methods. One is to compare citizen preferences with those of leaders and public policies. Much exploratory work of this kind has been done, but results frequently have been troubled by inadequate data concerning service delivery or other critical variables. Careful work is also necessary on methods for proper measurement of citizen preferences. By contrast, participation is relatively simpler to measure, but data are again sparse. Only one major study has surveyed participation in local elections, and it did not ask party preference. The "standard" participation model in empirical work includes only social background variables of individuals, although many theories stress issue characteristics, community conflict, and other contextual variables.

Knoke presents a particularly clear analysis of political culture, a subject which has not always received such treatment. He reviews past uses of concept, including ten basic elements. But the core of his chapter is a recasting of the typologies of Elazar and Banfield and Wilson. He shows that they essentially imply a continuum from pluralistic to unitary political culture. This continuum can then be applied to several political system characteristics, such as parties and legal structures. Results are summarized in his very suggestive Table 9.1, an extensive set of interrelated propositions specifying structural correlates of political culture. For example, cities with more pluralistic political cultures should have high turnout rates due to disciplined party effort, durable party cleavages along population lines, and few referenda, while cities with more unitary political cultures should be the opposite on these variables. Knoke then reviews several studies to assess empirical support for the propositions, although disparate data bases and analytical procedures complicate precise comparison.

Schumaker reviews ten central issues in studying citizen preferences and policy responsiveness. He argues that in assessing responsiveness it is not enough to study role dispositions of leaders or their perceptions of citizen preferences; nor are dyadic comparisons of single leaders and constituents sufficient. Actual policies, such as budget changes, should be compared with citizen preferences to capture system responsiveness. Similarly, participation by political activitists, as in voting, organizational membership, or contacting leaders, does not necessarily reflect the unarticulated concerns of average citizens. While census data, referenda, and simulations using national surveys have been used to assess local citizen preferences, direct survey data are more desirable. Such surveys should include not only controversial public issues but also noncontroversial issues and selected nonissues. A deductive list of issues can be formulated, Schumaker suggests, from considering generally important public policies. Collecting data for such issues for citizens, and specific responses by leaders, can indicate responsiveness to citizen preferences. Comparisons across cities differing in basic characteristics can illuminate structural variables.

POPULATION REDISTRIBUTION AND THE LOCATION OF ECONOMIC ACTIVITIES

Movement of population and location of economic activities have been subjects of considerable work. But some recent findings and new modes of analysis call into question earlier results, as reviewed in the Agenda. Standard analyses in this area build on the idea of central place theory—that people and firms seek to minimize their distance from a central place. Firms and residents compete for desirable land and bid up land prices, which leads to sorting out functions as distance from the central place increases. People are basically attracted to an area by their jobs and seek to minimize their travel distance between home and job.

These ideas have informed much research and found reasonable confirmation. But three types of new analyses question the results. One is more detailed study of specific job generation processes. Birch and others have shown that death rates or outmigration of firms differs less across cities than do birth rates of new firms, which are high in some growing suburbs and Sunbelt areas. Small- and medium-sized firms account for much of the growth, such that large firms which shift headquarters seem relatively unimportant. A second type of study is simultaneous equation analysis of interdependencies between jobs and population. These suggest the importance of the causal path from population to jobs. That is, people may move to attractive areas and, once there, create or find jobs for themselves. The tradition pattern—people moving to jobs—may still hold but should not be assumed the only, or perhaps even the most important, source of population movement. The third issue qualifying traditional work is the growing importance of public policies. Taxes and fees have increased in recent decades, as have services and other benefits provided by governments, which can shift location of population and firms. Interest in all these matters was particularly acute in congressional hearings in the late 1970s as various federal and urban programs were reviewed for their possible impacts on the rise of the Sunbelt and non-metropolitan areas and the decline of older cities. These three types of challenging results emerged from data that limit generalizability of the findings; addressing the same issues in the context of a study with more adequate data on the critical processes could help clarify these issues significantly.

Kasarda outlines some trends in the late 1970s heralded by some, such as increased middle-class location in central cities and "gentrification" of neighborhoods. The most adequate data to date suggest that such trends have not altered the more general patterns of urban population and economic base movement under way for several decades, but more careful analysis is needed to specify the magnitude of various shifts. Kasarda considers how to assemble a data file to address these issues. Such a file could be constructed, for the most part, from census sources, but these remain sufficiently complex to manipulate that assembling them in a single file over several decades demands rather specialized expertise. It is a task that could make basic data more accessible to younger researchers or nonspecialists in these matters. A file could be constructed involving all SMSAs, with data distinguished by central city and suburban rings, with subfiles included for cities and selected census tracts. The census data would be of general interest by themselves but would take on far greater interest if combined with data concerning political leadership, fiscal policy, service delivery, and citizen preferences. Some of these other types of data are available with varying degrees of completeness, comparability, and precision. And extending past efforts varies considerably in cost for the different components. If a regular data collection effort were mounted, elements could be added to it as funding and interest developed in specific areas.

Shlay and Rossi effectively illustrate the effects of population redistribution in their analysis of the impact of zoning on population and housing in the Chicago area. Pointing out that traditional central place theories, including human ecology, ignored political processes, they help restore this balance by introducing measures of zoning type for 395 central city and suburban tracts. Regressions initially estimate housing density, single-family, and multiple dwelling use for the tracts in 1960. Then 1960 to 1970 changes are analyzed, and zoning is shown to shift housing patterns as well as median family income and percentage of residents with families, although this occurs more in suburban then in central city tracts. However, zoning did not significantly affect the racial distribution across these tracts. The authors stress that the "natural" processes modeled by ignoring politics and government policies may have been more reasonable a half-century ago before zoning became an established element of American cities. But as times have changed, so, apparently, has the explanatory importance of different variables. For current and future population research, public-policy-related measures demand more explicit attention than they have received to date. Zoning is a particularly appealing variable, as it is relatively simple and inexpensive to measure. Nevertheless, if results are as dramatic as these with just one political process measure, imagine how much more powerful would be the combined effects of other measures, such as tax burdens and services.

In the Epilogue, Liebert examines the potential offered by a continuing data collection facility such as outlined in the Agenda and its implications for urban researchers, particularly the role the National Science Foundation has played, and could continue to play, in such an effort.

2

Urban Policy Analysis: A New Research Agenda

TERRY NICHOLS CLARK
RONALD S. BURT
LORNA CROWLEY FERGUSON
JOHN D. KASARDA
DAVID KNOKE
ROBERT L. LINEBERRY
ELINOR OSTROM

OVERVIEW

This is a proposal for a 12-month planning period to conduct conferences and develop specific plans for continuing work on comparative urban policy research. Two position papers would be commissioned for each of six conferences. The papers would review and critically assess existing data and specify major theoretical perspectives in the area. Some 15 urban researchers of diverse theoretical perspectives from the major social science disciplines would be invited to each conference. They would comment on the position papers and suggest specifically where and how they should be extended, as appropriate. At least two of the four project coordinators would attend each conference; then all would prepare a proposal incorporating as many suggestions as are practical for an integrated project on comparative urban policy research.

Enough important and overlapping work has been completed to date for several urban subfields to be considerably strengthened through a coordinated research effort. This might become a national resource project similar to the National Election Study, the General Social Surveys, the 1940-50 Census Project, and similar efforts in the natural sciences.

To date, a one-year grant from the National Science Foundation has permitted several informal meetings of urban researchers and a major conference in Chicago to consider these issues. The Chicago conference (held in April 1979) was attended by 20 urban researchers, most of whom presented papers that reviewed work in an urban subfield and suggested a series of specific next steps. There was a strong feeling among conference participants that we should seek to establish a continuing research project on urban questions.

The conference was summarized in a report which was distributed to several dozen other researchers. Comments were solicited in writing and pursued in discussions at professional meetings. These materials were then used by seven of the conference participants to develop this proposal.

The proposal begins with the rationale for the project, then discusses research to date and examples of questions to pursue in four interrelated areas. It next turns to organization for the activities over the 12-month period, and concludes with budget and background documentation.

RATIONALE

Most urban research has been undertaken by individual researchers or small teams using separate data sets. We certainly favor continuing individual projects, but the field has reached a point where strong arguments can also be made for a core of recurring data collection and analysis. Four basic arguments concern analytical cumulativeness, comparability of data units, longitudinal data, and economies of scale.

Analytical Cumulativeness. Probably the most compelling argument for the project is the greater analytical cumulativeness it could bring to urban research: If researchers have comparable data across cities and over time for several classes of variables, this can permit addressing questions in a far more systematic and cumulative manner than has been possible in the past.

If high-quality data become available, more researchers are likely to make use of them rather than of disparate samples and data sets sometimes unavailable to subsequent researchers. Replication, possible refutation, and extension of earlier work in a more cumulative manner are thus facilitated.

Dozens of disparate and sometimes contradictory results have been reported which might be resolved by more cumulative analysis. Just a few examples are the causes and consequences of centralization of decision-making (Clark, 1968; Grimes et al., 1976), the determinants of municipal fiscal policies (Bahl, 1968; Crecine, 1969; Weicher, 1970), variations in patterns of municipal service delivery (Lineberry, 1977), and the sources of citizen satisfaction with public services (Fowler, 1974; Rossi et al., 1976; Schumaker and Getter, 1977). In each of these areas several studies have been completed which sometimes support and other times seem at variance with one another. Measurement procedures and other aspects of the research have often differed. Comparable date concerning these issues could encourage more cumulative work.

Comparability of Data and Units of Analysis. Cities are more expensive to study than fruitflies or college sophomores. They also differ enough from each other that the particular types of cities studied can significantly influence results. Many individual studies have not had the time or resources necessary to collect data using a representative national sample of cities. These limitations of study design have influenced results to a degree that will remain unclear until more adequate data become available.

Longitudinal Data. Some results which apparently conflict may be due to differences in the time when the data were collected. There have been fundamental changes in cities over the last decades, especially federal policies affecting them, such as the expansion of categorical grants in the 1960s, block grants in the early 1970s, and increasing regulation and mandated programs in the mid- to late 1970s. Similarly, organized groups representing municipal employees and civic, business, racial, neighborhood, and environmental issues have periodically grown and declined. Basic shifts in jobs and population further alter tax bases and types of services which cities are called on to provide. A longitudinal study design which recurrently collects data on critical variables is essential to unravel the impacts of many interrelated variables.

Cost-Effectiveness. Almost all types of research have increased in cost with inflation. Large-scale urban research has always been expensive, such that only a few major urban studies have ever been undertaken as basic research. Most large urban studies have addressed specific policy concerns, such as racial disorders, productivity in urban service delivery, or evaluating federal programs such as Urban Renewal or General Revenue Sharing. Although data generated from these efforts have been of considerable importance for urban research, a number of critical data elements have often been omitted from any single study due to its specific focus. Currently, many individuals or small teams are pursuing studies which involve collection of urban data. Given the high costs of data collection, however, these individual projects are often limited in scope. This situation is quite understandable, since it is hard to justify a very large project for a single team. But then, given the disparate character of projects prepared by separate research teams, the data are often non-comparable and cannot be merged. This situation could change if a core set of data were collected and made available for urban research. It is more cost-effective to collect data for several overlapping studies in a single project. Further, once a major core set of data is collected and made available for urban research, individual projects could build on and extend it in various ways that would be responsive to the concerns of individual researchers, as well as more cost-effective for researchers, research agencies, and city officials, and citizens who provide the data.

THEORETICAL COHERENCE AND MULTIPLE DATA USERS

Given the scope of this kind of undertaking, it is important that numerous researchers of different perspectives have an opportunity to help shape the effort such that it can generate results useful for their research. This proposal thus requests support for a planning phase structured around a series of small conferences to permit numerous researchers to provide input to the project. These will permit dissemination of information about the project to interested researchers, help assemble a number of sets of research hypotheses, assess alternative data collection procedures, locate existing data sets, and so on. Specifics are discussed below.

It is much easier to agree on data to collect than to reach theoretical agreement. But theoretical agreement is unnecessary as long as researchers are close enough in their hypotheses that they can use the same basic data. Therefore, neither this proposal nor the planning conferences are intended to generate an integrated set of hypotheses of the kind one would expect from a single research team. Subsequent sections list a number of competing and complementary approaches and propositions that would make this document inevitably more open in focus than a normal statement from a single team. Although only seven of us were actively involved in its writing, we have sought to reflect the concerns of many other urban researchers.

Still, it should also be emphasized that there is no lack of coherent theory. The research traditions reviewed in the next section are above average in their degree of coherence for the social sciences. Indeed, we have sought to include certain lines of research which fit together and where the work to date is strong enough both to inform specific research hypotheses and to benefit significantly from a national urban data collection effort.

Basic Research and Public Policy Concerns. Most urban research involves both basic research and more policy-oriented work. Often the two overlap considerably. As this proposal is being submitted to the Division of Social and Economic Sciences of the National Science Foundation, we do not seek to spell out many policy implications, although the interested reader can see many just below the surface. The project has been developed with a predominantly basic science emphasis, but we nevertheless feel that it would benefit from involvement of persons directly concerned with urban policy questions—city officials, federal and state agency staff, public interest groups, and private participants involved with such questions. We are in contact with several persons who have such policy-related responsibilities and who are conversant with problems of more basic research; they might be informed of our activities, participate in the conferences discussed below, and remain involved on a regular basis as the project proceeds.

REVIEW PROCEDURES AND FUNDING

The scope of the project is such that after the planning phase it would obviously surpass the normal level of a single-team study. Thus we expect it to be reviewed by procedures appropriate for such large projects, more common in the natural sciences than the social and economic sciences. The proposal has been undertaken with the expectation that it could bring generally enhanced support for urban research without necessarily diminishing the normal funding sources for such work. We make the point here as it has been raised by urban social scientists less familiar with large projects of this sort and who may understandably raise it in the review process.

A FEW GENERAL REMARKS ABOUT URBAN RESEARCH

Like many fields in the social sciences, urban research has been informed by several general theoretical perspectives, including systems analysis, functionalism, neo-classical economics, the new political economy or public choice, and neo-Marxist writings. But as most of these general theories have little specifically urban content, they do not directly inform much empirical research. Few urban researchers are committed to any single general theory; most draw selectively from more than one. Rather, most theory in the urban area consists of middle-level proposition, clusters of theory, and empirical results which help explain some major processes affecting cities: the sources of migration to cities, of city government expenditure levels, and so on. As these subfields have often been defined by empirical processes that affect cities, they have attracted researchers from different disciplines. Much urban research thus has an interdisciplinary character, often involving significant contributions from political scientists, sociologists, economists, and geographers, among others. Specialists on migration, for example, may share more concerns with fellow specialists from adjoining disciplines than they do with disciplinary colleagues working in other areas.

Much urban research has significant policy implications, ranging from such issues as how to improve garbage collection procedures to indicating how many cities are fiscally strained. Several major data collection efforts have been undertaken to address pressing policy issues; these efforts have provided several researchers with experience in such undertakings, even if time pressures and specific policy questions have limited the generalizability of some such activities.

Still, most urban research is conducted by university-based social scientists who spend much of their time building a cumulative body of knowledge concerning urban phenomena. While the most ambitious data collection efforts to date have been funded by policy-oriented agencies (especially NSF-RANN and HUD), much continuing data analysis is underwritten by small grants and university contributions of faculty time, research assistance, and computer and library facilities. Many urban researchers thus move back and forth between more policy-oriented and more basic science activities; each has clearly reinforced the other.

There is fairly general recognition of the empirical interpenetration of some major urban processes. Migration patterns in and out of cities alter their social composition and tax bases, which in turn affect patterns of political leadership, government policies, and citizen satisfaction. But while the interpenetration of such processes has been partially codified in many propositions, there have been few efforts to construct formal models involving these varied processes.

BASIC URBAN PROCESSES: FOUR FOCI FOR RESEARCH

It is impossible to study everything urban. This point has been made repeatedly in discussions we have had with various researchers concerning the project, and was underlined by participants at the Chicago conference. It follows that a limited number of processes should be selected and investigated in detail. While some arbitrariness inevitably influences such selection, at the Chicago conference and in preparing this proposal an attempt was made to distinguish basic processes which are (1) particularly important in their effects on many American cities and of continuing concern to citizens and policy makers, (2) already studied enough that some codified results exist to guide future efforts, (3) sufficiently interrelated to justify a joint research undertaking, and (4) amenable to data collection strategies at reasonable costs.

Researchers of distinct theoretical perspectives are engaged in work on each process. While cooperative research activity can generate theoretical exchanges and heightened awareness of conceptual alternatives, we in no way assume that a unitary theoretical perspective is necessary. Rather, a common body of core data that can be used to address a range of theoretical questions seems the appropriate goal for an enterprise of this kind. In this spirit, we have sought to indicate core propositions and research questions shared by many researchers. Four processes reflecting these criteria are outlined briefly and discussed further below.

Population Redistribution and the Location of Economic Activities. These include urbanization patterns involving migration from rural to urban areas; deconcentration of population to suburbs and outlying areas; interregional movements, such as the rise of the "Sunbelt" and decline of the "Frostbelt"; and more limited movements generating decline or "gentrification" of neighborhoods. Both people and jobs move in such a manner that the socioeconomic base of urban areas can change, often quite rapidly. Literatures on these specific questions have developed in several of the social sciences. These phenomena are important both in their own right and in setting a context for the three other processes.

Political Leadership and Fiscal Policy-Making. Changes from migration in the size of the various income, occupational, and ethnic groups of citizens feed into the political process. Inputs may be direct in the form of votes for new political leaders, or more indirect—mediated, for example, by networks of organized neighborhood groups, political parties, and business groups. These issues are addressed by literatures on community power, political culture, public choice, coalition theory, network analysis, and budgetary theory. Fiscal policies are the major outputs of this process; they in turn constitute inputs to the next.

Bureaucratic Processes and Service Delivery. Political leaders who vote a budget still do not deliver services to citizens. In between are municipal employees who staff agencies; how they implement specific policies depends on a number of characteristics of the personnel, agency structure, and units being served. A traditional public administration literature on such issues has been extended in recent years by efforts of many city staffs as well as researchers to develop specific performance indicators which can help unravel the determinants of productivity in service delivery.

Citizens: Preferences, Participation, and Community Contexts. Services delivered to citizens partially inform their views of what their govern-

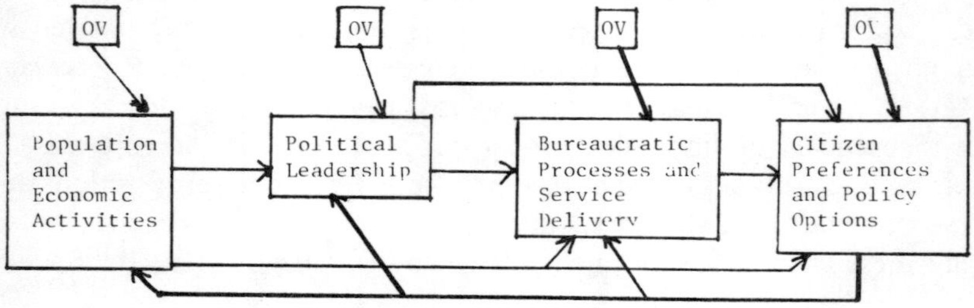

Figure 2.1 A Basic Framework of Variables
OV = other variables

ments are doing. But citizens differ considerably in their expectations concerning the quality of public services, taxes, and other charges they are ready to pay, and conceptions about more general process of urban decision-making. Such differences in orientation depend partly on socio-economic characteristics of individual citizens, as well as city-specific variables (tax burdens, service delivery patterns, and the like.) Research on political behavior, citizen preferences, and participation informs these issues.

Figure 2.1 portrays a basic input-throughput-output model of the four basic processes considered here. Similar models have been presented in many research efforts in this area. If the causal thrust of most analysis has moved from left to right among the variables shown in Figure 2.1, there are also feedbacks, indicated by the arrows pointing back from right to left. Similarly, "other variables" not explicitly included in this framework affect each variable. Developing precise empirical estimates of such complex relationships involving feedbacks demands comparable data for many cities over time. Few urban research efforts have achieved this kind of analysis; we hope to change this.

We next turn to a more detailed consideration of work in the four general areas. Each includes a brief review of some major lines of work to date. Our concern is not to develop specific theories or propositions in any detail, but to indicate a range of approaches dealing with similar questions. Then specific variables and data important in the context of several theories are listed. The point of this is to help provide an agenda for organizing the conferences which is open enough to encourage discussion but specific enough to provide focus.

POLITICAL LEADERSHIP AND FISCAL POLICY-MAKING

THEORETICAL APPROACHES

This section briefly reviews some major theoretical approaches to political leadership and fiscal policy-making. It then indicates a series of variables these approaches point to for empirical analysis.

Elite Theories. A long tradition of elite theories has stressed the critical role of distinctive leaders. Writings of Marx, Weber, Mosca, Pareto, Michels, and others informed an empirical approach to urban leadership (see Hunter, 1953). This tradition stressed the *social and occupational background characteristics* of leaders, to assess the degree to which various occupational, educational, income, and ethnic groups were involved as leaders. *Reputational power* of leaders was tapped using Hunter's "reputational" methodology. *Social networks* among leaders were recorded using the sociometric procedures of the day.

The emphasis on social background characteristics and institutional arrangements has been elaborated by a neo-Marxist tradition, which emphasized specific characteristics of the *economic base,* the degree of countervailing *organized opposition to business elites* by *poor* or *workers* (private or public), and the increasing role of the welfare state as means of legitimation (see O'Connor, 1973; Harvey, 1973).

Network Analysis has studied linkages among leaders using more refined methodogies. Separate *social, economic, and political networks* are distinguished, and patterns of values, social, and economic characteristics of leaders compared (see Laumann and Pappi, 1976; Marsden and Laumann, 1977; Galaskiewicz, 1979; Burt, 1977; Brieger, 1979). Such networks may link important individuals, but they also provide a convenient way to analyze interorganizational relations (Turk, 1977).

Organized Groups. The tradition of group theory from Bentley, Key, Truman, and Dahl stresses competition among interest groups as the primary source of policy outputs. Dahl (1961) emphasized the distinction between *resources* and *influence,* as measured by *participation in decisions.* Similarly, he stressed the importance of studying specific *issue areas* and considering their differences in terms of decision-making patterns. Patterns by which specific organized groups may mobilize their resources more effectively to influence public policy have been a concern of many others (for example, Clark, 1968; Gamson, 1968; Coleman, 1971).

Coalition Theories concerning the number and types of specific groups likely to join one another in exercising leadership have led to numerous hypotheses. The basic size principle postulating a minimal winning coalition (Riker, 1962) has been complemented by propositions about attributes of the participants, resource distributions, ideological distance, and other characteristics (Gamson, 1964; Groennings et al., 1970; Swaan, 1973; Hinckley, 1979). In particular, policies provided by governments have been classified as public or separable goods. Free rider problems involving support for public goods can be solved by simultaneously providing separable goods, or via the activities of "political entrepreneurs" (Olson, 1965; Frohlich et al., 1971).

Citizen and Leader Preferences have been stressed in work since Downs (1957) using median voter models. Under conditions approximating those of a perfect market, political leaders should become "invisible" and supply public policies to citizens, just as a private entrepreneur supplies private goods. But numerous imperfections in such market operation make introduction of other variables important. Most formulations thus stress the importance of comparing preferences of citizens with policies proposed or provided by leaders. The degree of disparity between citizen preferences and policies provided has been analyzed theoretically as a "loss function" (see Riker and Ordeshook, 1973) and empirically as "policy responsiveness" (see Schumaker and Getter, 1977). One proposition from these traditions is the greater the disparity between the preferences of citizens and existing policies, the more likely are citizens to become politically active.

Political Culture has been used with two basic referents: the more general sociocultural context, which helps define the rules of the political game, and the symbolic policies, which characterize local leadership. Both have been classified along a continuum from unitary or public goods to more pluralistic or separable goods. These in turn have been linked to ethnic, religious, and socioeconomic characteristics of the population and political leaders (Wilson and Banfield, 1964; Elazar, 1966; Williams and Adrian, 1963; Clark, 1975a), as discussed further in the section on citizens. Other culturally linked value patterns include local autonomy, citizen participation, and conflict-avoidance (for example, Jacob et al., 1971; Patterson, 1968).

Bureaucratic Theories by incrementalists like Meltsner (1971) and Crecine (1969) have sometimes seen bureaucrats as sources of public policy, in the passive sense of respecting bureaucratic rules. By contrast, a more active role is assigned to bureau heads who seek to expand their staffs and budgets in work by Niskanen (1971) and others (such as Borcherding, 1977, and Friedman, 1975).

Fiscal Policy is a product or output of the political system explained by each of the seven theoretical perspectives just listed. But while these seven have developed as distinctive, if overlapping, perspectives to explain fiscal policy, they have seldom been systematically confronted as alternative interpretations, nor have data been analyzed to assess their appropriateness. Two considerations may help explain the relative explanatory power. First are contextual variations across cities at one point in time. For example, a small, homogeneous suburb is more likely to have a clearer and more distinctive set of citizen preferences and leaders who implement them. By contrast, in a large, socially heterogeneous city, more complex relations among organized groups and elites may generate shifting coalitions and fiscal policies. Such variations may be partly analyzed using cross-sectional comparisons of cities, as well as cross-sectional comparisons of changes in fiscal policy, given appropriate data. Comparison of properly selected individual cases can also be helpful. Second, these patterns vary over time because of demographic, economic, and other changes that alter the makeup of the city, but also because of new federal programs such as categorical grants of the Great Society years, block grants of the New Federalism, and continuing changes in regulation and administrative procedures affecting local decision-making and fiscal policy. Some observers feel that the impact of these developments has been not just in the formal structure of intergovernmental relations; the content of local decision-making has also been fundamentally altered. Have the cumulative effects of federal programs been to shift autonomy from city halls to higher levels of government? Why has there been substantial growth in some local governments but not others? What local factors encourage taxpayers' revolts and fiscal retrenchment? And where and why are significant efforts undertaken to improve productivity? Such issues are debated in national urban policy circles each time a new urban program is considered. The principal evidence remains testimony by local public officials, although in recent years the U.S. Conference of Mayors, members of Congress, and federal administrators, as well as social scientists, have sought more systematic data to address these issues.

Despite some important efforts that have concentrated primarily on individual programs (for example, NSF-RANN, 1976; Nathan et al., 1977) we only have partial data to assess such general questions. Collecting data for a national sample of cities over several years is critical both for adequate policy assessment and consolidation of alternative theoretical perspectives on urban political leadership and fiscal policy-making. This conclusion was articulated at our Chicago conference, as well as at a 1979 session of the American Political Science Association on theories of urban politics (including four papers and two discussants of quite varied theoretical perspectives). In brief, we have a reasonably clear sense of the major variables affecting leadership and fiscal policy and the subprocess by which they operate; but how they fit together and differ across cities and over time will remain unclear without better data.

SURVEYING URBAN LEADERS

Although these theoretical perspectives are numerous and diverse, they generally imply collecting data from elected and administrative officials in city government and leaders of important local organized groups. Leaders themselves, as well as knowledgeable informants about them, are likely candidates for interviews. Although we often refer to leaders in this section, in fact many aspects of leadership take on meaning largely by virtue of the leaders' organizational positions. Even if many data items are collected from individuals, they can be analyzed with organizations as the focus.

A large amount of research has accumulated surveying urban leaders, including results for several items that can provide baseline data for comparison with future research (see Clark, 1975b; Laumann and Pappi, 1976; Caputo and Cole, 1977; Kraemer and Dutton, 1973; Jacob et al., 1971; Schumaker and Getter, 1977; Eulau and Prewitt, 1973).

New survey procedures (telephone interviews, direct data entry, and the like) have reduced costs of surveys generally and make surveys of leaders in numerous cities less expensive than was the case a few years ago, despite the general increase in research costs. Similarly, research to date has suggested that reasonably reliable results can be obtained from a dozen or even fewer informants in a city. These cost factors, combined with experience concerning some types of items that work reasonably well (and others that do not), suggest that it might be feasible to aim for a continuing survey of leaders in 50 to 100 cities perhaps every two years. Data of this kind would generate an invaluable time series for numerous research and policy purposes.

To articulate the political leadership survey with service delivery patterns and citizen preferences, it is important to focus on a limited number of important service areas. Given the complexities of studying services provided by special districts and other intergovernmental arrangements, it seems reasonable—at least at the outset—to focus on municipally provided services. The following set of considerations led us to six possible service areas.

Cities produce both goods and services. Goods involve the transformation of physical inputs into physical outputs, while services involve changes in the personal characteristics or attributes of consumers (Garn et al., 1976:16; Fuchs, 1973). Goods can be produced independently of the participation of consumers and can be measured regardless of their effect on consumers. Services necessarily involve consumers as *part* of the production process itself if there is to be any production of the service at all. Measurement of the amount of services produced involves the measurement of use patterns. Municipal facilities (such as public housing and libraries) have attributes of being both goods and services. The construction of the facility itself is a physical transformation and does not directly affect consumers, but use of the facility depends on the active participation of consumers.

Including two examples of each of these three types of municipal goods and services would lead to the selection of six service areas (or functions) for intensive study. As a preliminary list we propose to include road repair and maintenance and solid waste collection as the two physical products; police as a critical service; publicly assisted housing and libraries as municipal facilities; and urban development as a combination involving zoning, tax incentives, and other policies concerned to affect the population and economic base. These six service areas also span the range from relatively separable goods to relatively pure public goods. They include services which are highly salient to most urban residents and those which may dramatically affect the spatial location of population and economic activities, leadership patterns, and citizen participation. Final decisions on the services to be included should await consultation with researchers in the area, but this is offered as a reasonable initial list.

Leadership Characteristics. These involve general characteristics of leaders and their organizations. Basic organizations and positions studied in work to date and which tend to be important in most American cities include the following (variables for which data might be collected are indented):

Public employee associations and unions (city, metropolitan, state)

Political parties

Religious organizations

Newspapers, radio, and television

Business organizations (retail merchants, Chamber of Commerce, bankers, industrialists)

Professional organizations (Bar Association, Medical Association)

Civic and charitable groups (League of Women Voters, service organizations such as Rotary, Kiwanis)

Neighborhood groups

Environmental groups

Taxpayer associations

Elderly groups

Ethnic or racial organizations

Private sector unions

Mayor

City council

City manager (if applicable)

Heads of local government agencies

Certain data may be collected for many or all such organizations and positions by interviewing several leaders as informants. Resources and reputational power are two basic leadership characteristics.

Resources (such as budget of the organization; staff size; and, if it is a membership organization, number and social characteristics of members.)

Reputational power (measured with "issue-specific reputational" items or similar formats, such as "how important is each of the following to be elected mayor").

Social background characteristics of the leader (age, sex, race, education, occupation, party, brief career history, and so on).

Participation. This is important for elected officials, organized groups, and citizens. The leaders interviewed can serve as informants about a range of organizations and provide more detail about their own involvement.

Influence or participation in actual decision-making (measured using "decisional" items such as who initiated, supported, opposed, mediated, and prevailed in various specific decisions).

In such questions, as in those involving reputational power, it is common to list several specific decisions that are either identical or similar in each city, such as the last mayoral election or the largest single decision in last year's budget. A longer, more elaborate version using the same general format can be used for important individual leaders, referring to themselves or their organization.

Similar items can also be used to assess the participation of various types of citizens in the same decisions. It is interesting to know how responsive elected officials in particular are to major types of citizens, and to compare responsiveness as reported by leaders with results from a citizen survey, as discussed below.

Policy Preferences and Political Culture. These two areas are also important for citizens. They are discussed later.

Social Networks. Several of the above types of items may be used to construct social networks: social characteristics of leaders and organizations, reputational power, and participation in various decisions. Similarly, questions about interrelationships among individuals and organizations can be posed. (1) Researchers could inquire about direct contacts (such as how often one leader or organization has contacts of different kinds with several other specific organizations). (2) Network data might be collected in much greater detail in perhaps four cities. The four could be selected partly on the basis of the more brief network items used in all cities. Burt's (1979) conference paper outlines this methodology.

Bureaucratic Processes and Service Delivery. Questions in this area are discussed in the section on bureaucratic processes.

Fiscal Policy Outputs. Certain data concerning general fiscal policies (general revenues, debt, general expenditures) are readily available from the Census Bureau's *City Government Finances* series. Several modifications of these data are in order. Detailed breakdowns of intergovernmental revenue, distinguished by types of federal and state sources, are being prepared by Thomas Anton and his associates at the University of Michigan. These could be added to the project data files in consultation with him.

Some other weaknesses of census data can be overcome by rather simple recoding from documents and worksheets at the Washington office of the Governments Division. For example, short-term debt can be distinguished as to whether or not it is guaranteed by the federal government (as for many housing and urban renewal projects). It may also be worthwhile to code changes in functional responsibility over time using the Governments Division worksheets.

Fiscal data from local sources can also be coded, and these may be necessary for assessing certain past changes in function or determining certain specific current functional responsibilities (such as ambulance service, inoculation programs, and fingerprint detection). Local data may also have to be consulted to permit comparable allocation of expenditures to major services (for example, by assigning pensions and overhead in a comparable manner).

BUREAUCRATIC PROCESSES AND SERVICE DELIVERY

In comparative urban research, service delivery has been viewed as an independent, mediating, and dependent set of variables. A study designed to provide useful data to a variety of urban scholars needs to collect appropriate data for viewing service delivery from any of these three perspectives. When considered as an independent set of variables, urban public goods and services are seen as one of the inputs into location decisions made by industrial and commercial enterprises and individual households (Hawley, 1957; Kasarda, 1972; Neenan, 1970; Tiebout, 1956). Thus, to understand the process of spatial location and economic activities, one needs good estimates of the quantity and quality of urban goods and services available. Further, one needs to ask key questions about the distribution of these goods. Are the goods distributed primarily to the central business district, to industrial users, or to residential neighborhoods? Do residents living outside the jurisdiction receive a large proportion of the goods and services produced? Are the more affluent neighborhoods the recipients of higher levels of urban services than poorer neighborhoods? Are racial or ethnic characteristics of neighborhoods related to distribution patterns?

In addition to the effect on location decisions (entry and exit decision), the quantity and quality of urban services can also be viewed as an input to the decisions of citizens to participate or not participate in local politics (voice; see Hirschman, 1970; Orbell and Uno, 1972). While the factors which affect citizen participation are numerous, dissatisfaction with services has been shown to be related not only to voting behavior but also to a variety of other forms of citizen participation, including contacting local public officials. Satisfaction with services appears to be related less to actual quantities and qualities available than to the relationship between the type of service provided and the preferences of citizens for different types of services. Thus, to understand some aspects of participation, one needs to gain knowledge of service levels and the responsiveness of service delivery patterns to citizen preferences.

The number of urban services produced by a municipal government has also been viewed as a key mediating variable. The accessibility of leaders, level of citizen participation, program innovation, amount of expenditure, and accountability are viewed as affected by the number of urban services produced by a municipality, as well as by the age of a city, its size, and whether it is a central city or a suburb (Liebert, 1976). Cities which produce a larger number of urban services are seen as affecting a broader scope of community affairs and providing a greater focal point for the resolution of key community conflicts. The fragmented nature of the service provision in newer municipalities has been viewed as a pathology needing to be corrected by one or another form of metropolitan reform (Committee for Economic Development, 1970; Erie et al., 1972; Friesema, 1966). Consequenly, measures of diverse arrangements for production and provision (Ostrom et al., 1961) of different urban public goods and services are needed as mediating variables to understand patterns of urban leadership, citizen participation, conflict resolution, and innovation.

The processes of urban service delivery have themselves been the object of considerable empirical research by comparative urban researchers. The theoretical traditions of researchers in the field vary widely. However, almost all students of urban service delivery consider the understanding of bureaucratic processes as essential for the eventual understanding of the process of urban service delivery more generally. Researchers who examine urban service delivery from an organization theory perspective have focused on structural differentiation, span of control, informal organization, and incentive systems of bureaucracies producing services. The incrementalist literature stresses the burden of established routines in structuring future activities. The self-interested bureau chief is emphasized as a source of internal policies in other writings. The technical efficiency of different production functions adopted by bureaucracies has been the focus of many studies. The interaction of bureaucratic structure with the type of public goods produced has been the focus of other scholars. The impacts of professionalization, civil service, and unions on bureaucratic decision-making have also been of major interest to some in this field.

However, for all of the diversity of approaches, a high level of agreement exists among specialists of bureaucratic processes and urban service delivery on (1) the need for a large, urban data base containing valid and reliable measures of the key variables involved in urban service delivery and (2) what these key variables would be. Major questions of theoretical and policy import cannot currently be addressed in the context of a large number of diverse urban settings with the available data. Valiant efforts have been made to undertake secondary analysis of data collected for other purposes. However, when data available for secondary analysis have been subjected to careful examination, and comparisons made to validate data, major problems of validity and reliability have repeatedly been discovered.

Expenditure data collected by the U.S. Bureau of the Census have been frequently used in comparative urban research to measure effort devoted by a city to the provision of particular types of services. In a critique of the simple acceptance of expenditure data, Liebert (1974) demonstrated that one needed to know the particular services actually being produced by a city (and thus included in their expenditures) prior to undertaking this type of analysis. Whether or not a city itself produced education, water, sewers, electricity, general assistance, welfare or public health dramatically affects both its total expenditure levels and the proportion of expenditures that is devoted to any particular type of service. Once data on the number of urban services produced by a city were included in analysis, previously established findings about the relationships between the structure of local government and expenditure patterns were seriously challenged.

Recent research in the area of policing has shown that it is necessary to know not only about the specific municipal producers, but also about the presence or absence of other major producers of a service within the same city (Ostrom et al., 1978). If two cities produce police services for the same number of residents but a major military base and a major university with large police departments of their own are located within one of those cities, per capita expenditures for the two cities cannot be validly compared without controlling for the expenditures of the enclave producers. Without adequate data about the full range of production arrangements within a city, grossly incorrect inferences about service delivery patterns will be made.

FBI crime statistics have been used frequently in comparative urban research as a measure of police output (Chapman et al., 1975; Phillips and Votey, 1977). While this practice has been subjected to serious criticism on theoretical grounds (Biderman, 1966), recent comparisons of the victimization data collected by the U.S. Bureau of the Census for LEAA have demonstrated that inferences derived from analysis of FBI crime statistics are significantly different from inferences derived from analysis of the LEAA victimization data (Booth et al., 1977). Problems in the use of municipal expenditure data, production arrangement data, and the FBI crime data are but three examples of the many extreme problems facing urban researchers wishing to undertake serious analysis of bureaucratic processes and urban service delivery in a large number of cities using currently available data.

While progress at the level of broad generalizations resulting from the study of bureaucratic processes and urban service delivery for different services in many locations has been stymied by lack of good data, significant progress has been made in challenging old myths and coming to new (albeit tentative) substantive understanding of the types of bureaucratic processes which affect urban service delivery in specific locations. In a series of small, comparative studies within metropolitan areas, previously accepted notions concerning the efficiency of large-scale urban police agencies have been seriously challenged (Ostrom et al., 1973; Ostrom and Parks, 1973; Rodgers and Lipsey, 1974). The production strategies adopted by large agencies dramatically differed from those of smaller and moderate-sized agencies. While small to medium-sized police agencies allocated a larger proportion of their officers to on-the-street duty and stressed crime prevention and service activities, large police agencies allocated a higher proportion of personnel to specialized units located in headquarters and stressed postvictimization follow-up. Victimization rates were lower and speed of response and citizen evaluation of police were higher in the small and middle-sized jurisdictions.

Some initial studies of contracting by city government with private firms for such services as fire protection (Ahlbrandt, 1973) and solid waste disposal (Savas, 1977) provide important insight into the level of innovation and alternative patterns of personnel assignment and utilization adopted in private firms as compared with public agencies. The same internal processes may not operate when smaller cities contract with large-scale public agencies if the findings from a study of contracting conducted in the Los Angeles metropolitan area (Sonenblum et al., 1977) are verified in other areas.

The widely accepted "underclass hypothesis" that urban services are most frequently distributed in an unequal manner favoring the wealthy, white neighborhoods and the powerful (Lineberry, 1977) has not fared uniformly well in a number of multineighborhood, single-city case studies conducted in Oakland (Levy et al., 1974), San Antonio (Lineberry, 1977), Houston (Mladenka and Hill, 1977), and Detroit (Jones et al., forthcoming). In all of these well-constructed studies, internal bureaucratic decision rules have been the most powerful factor affecting how urban services were delivered. Bureaucratic decision rules concerning the allocation of park and library facilities based on a population-related measure appear to produce relatively equal distribution of these facilities in most of the cities studied. However, bureaucratic decision rules concerning the allocation of budgets for the purchase of books based on previous use patterns appear to lead to an unequal distribution of funds for book purchases to middle-income and wealthy neighborhood libraries. This inequality is probably exaggerated still further by the bureaucratic decision rule that the type of books purchased for all branch libraries in a system should be similar. Residents in a poor neighborhood might utilize their library more intensively if materials more closely related to their concerns were included. The apparent equality in the type of books purchased seems to lead to inequality in use patterns, which in turn is reflected in inequality in budgetary allocations for book purchases.

A recent intensive study of police response time conducted in Kansas City (Kansas City Police Department, 1977) found that the average time for a citizen to report a crime to police is greater (by far) than either the time (1) to relay the citizen-reported information through the dispatcher to the patrol unit or (2) the travel time of the patrol unit to the incident. The researchers also found that faster *citizen* reporting times were related to a higher proportion of arrests upon arrival and to the location of witnesses. The importance of citizen activities was also stressed in a study of criminal investigation conducted by Rand (Greenwood and Peterselia, 1975). Fauri (1978) has stressed the importance of citizen involvement in all human service delivery. Beyond bureaucratic processes, the extent of citizen involvement as active coproducers of urban services should be assessed in future research.

In addition to major progress in substantive understanding of urban service delivery in particular locations, major breakthroughs in data collection methodology have also occurred during the past few years. In all the specific studies citied above, careful efforts have been made to develop and test new indicators for both service activities and outputs and to disaggregate these to the neighborhood level. Harry Hatry and others at the Urban Institute have devoted years to the development of a number of valid and reliable instruments for collecting data about street cleanliness (Blair and Schwartz, 1972), street repair (Boots et al., 1972), recreation (Winnie and Hatry, 1972), and citizen surveys for measuring service performance (Webb and Hatry, 1973). Elinor Ostrom, Roger Parks, and their colleagues developed several different modes of data collection for street lighting and road repair and examined relationships among the different ways of measuring these two services. Valid ways to measure production arrangements in metropolitan areas have been developed for police (Ostrom et al., 1978), fire (MacGillvray et al., 1977) and solid waste disposal (Savas, 1977).

Without this recent substantive and methodological progress, a major investment in new data collection would not be warranted. However, the area of bureaucratic processes and urban service delivery is now ripe for a major investment for the three reasons briefly discussed above: (1) Little progress in broad generalization and cumulative theory development can be made at this time without a major investment in new data collection in a number of different locations. (2) Studies conducted in a limited number of cities have produced substantively important findings on how bureaucratic decision processes affect urban service delivery. (3) The methods to ensure valid and reliable data collection efforts are more fully developed at this time than ever before.

SPECIFIC VARIABLES FOR POSSIBLE STUDY

Variables from several different sources can affect bureaucratic processes and service delivery, but here we list primarily those which seem most directly related to these matters. As discussed in the political leadership section, six services might be studied, and some data collected for each, via the (1) service agency, (2) service agency chief, (3) political leaders, (4) citizens, (5) block faces.

(1) Service Agency Data. Data could be collected for each of the six service areas, possibly using a coding form to be completed for the agency which is the primary producer of the service for the sampled city. This is usually but not always a city department. Variables could include:

Type of organization (city agency, special district, private firm, other)

Finance (what are the sources of funding for this service; detailed expenditure information for noncapital expenditures for past two years and for capital expenditures for past five years)

Personnel allocation (number of authorized FTE, number of current line personnel by division, number of current staff personnel by division, number of part-time staff and estimate of total hours, number of auxiliary personnel and estimate of total hours, number and type of personnel employed under special federal programs during past five years, total personnel for past five years)

Civil service regulations (when adopted if present, type of entry-level examination, type of promotional examination, who covered, who prepares and grades examination, how much discretion is left to agency chief)

Unionization (which employees, if any, are unionized; which unions; agreements include salary, working conditions, or both; length of union contracts)

Salary levels for entry, mid-range, and top-level line and staff

Entry requirements (residence, education, training immediately after entry)

Organizational arrangements (how is executive selected; if selected, how frequently; presence of special governing board; presence of special citizen review board; number of levels of hierarchy; span of control for street-level supervisors; span of control for highest executive)

Organizational diversity (number of separate services produced, number of separate occupational titles, number of major organizational divisions)

Innovation (recent reorganizations in way agency is organized, recent additions of new services, recent adaptations of new production technology)

Level of demand (total calls for police service, average daily patronage at library, number of zoning variances filed, etc.)

Activity indicators (number of miles patrolled, number of hours facilities available to public, number of vehicles used simultaneously)

Output indicators (number of arrests, number of crimes cleared by arrest, library circulation levels, tons of garbage collected, miles of roads maintained, etc.)

Internal performance measurement system (how does agency decide how good a job is being done, any performance measurement system for employees)

For those cities in which a sample of neighborhoods has been selected, additional service-specific information could be collected at each agency but coded for each neighborhood.

(2) Service Agency Administrators. While most of the above data about service agencies could be collected from lower-level officials, some key information is available only from top-level administrators.

Budgetary process (who within the agency is involved in initial budget projections, how is budget prepared for agency, who reviews for the mayor, what type of budgetary review is made by the council, what type of expenditure review is make by mayor and council after apporpriations, how are decisions made concerning allocations to particular neighborhoods)

Political input (frequency and type of involvement with mayor, manager, specialized governing board members of the city council, other city agencies; citywide organizations or party organizations such as Chamber of Commerce, League of Women Voters; service organizations such as Rotarians, neighborhood organizations, individual citizens)

Professional associations (frequency and type of interaction with other agency heads in metropolitan area, in state organization, in national organizations; membership in professional association; officeholding in professional associations)

Performance measures (which agency-wide performance measures does chief consider important, how are they performing compared to national standards, how frequently are internal management reports prepared, what type of data are contained in internal reports, what type of performance indicators are used for individual performance evaluation)

Federal and state involvement (how does agency head view federal and state role in financing or regulation of service production, how involved is the agency with federal or state programs, what consequences for agency, etc.)

Role of metropolitan-wide coordinating efforts (if any coordinating mechanism related to this service area, how do these affect agency production practices; any joint activities, any purchasing agreements, any conflict between individual agencies and metropolitan coordinating mechanism)

Conflict and/or cooperation among agency (what patterns of conflict and/or cooperation have occurred between parallel production agencies in the metropolitan area)

Internal decision-making (what types of operational decisions are made at different levels within agency, what standard operating practices are used, etc.)

Personal characteristics (age, residence, sex, party affiliation, education, profession, etc.)

Employment history (when appointed or elected to this position, if elected, how many times reelected; prior association with this agency; prior work history)

(3) Political Leaders. Most questions on this form would come from the political leadership section. Items specific to the six service areas that might also be included are:

Evaluation of the general quality of each service in the city (is the level of output and quality higher, about the same, or lower than other cities in metropolitan area and other cities of about the same size; what indicators are used for evaluation of service quality and quantity)

Distribution patterns (do they feel that each service is distributed in an equitable fashion within the city, what type of inequities might exist)

Citizen communication concerning service (how frequently do citizens contact political leader concerning each service area, what is the type of request made by citizens, does contact come more from organized groups than individual citizens)

Communications with agency (how frequently does political leader contact agency, who in agency is contacted, what type of problems or questions are involved).

(4) Citizen Data. These items are discussed in the citizen section below.

(5) Block Face Data. For a sample of block faces in each sample neighborhood, carefully trained observers could be sent to complete a detailed coding form. The form would draw on similar forms developed at the Urban Institute and Indiana University. The coder would first make a sketch of the block face in the sample and the block face opposite the one sample. Data would be posted to a structured coding form but also sketched onto the drawing. Types of specific variable might include:

Block length in feet

Number and type of housing units, number of commercial and industrial units

Number of set back and other obvious zoning infractions

Number of abandoned residential, commercial, or industrial units and their condition

Number, type, and location of street lights

Number and location of traffic lights or stop signs

Number of "For Sale" signs

Street cleanliness level

Road roughness level

Sidewalk cleanliness level

Amount of graffiti on buildings

Weed level in vacant lot or curbside

For single family residences, number of chain-link or other protection-type fences

Number of large dogs visible from sidewalk

Number of night lights visible from sidewalk

Number of persons walking on street while observer present (standardized for length of time used for observation and number of housing units)

Number of persons using outside porches (standardized in above manner)

Did anyone question why observer is present and making notes

If observation during local election, count number of local election signs

Code from observation and maps distance to nearest library, fire station, bus stop, park, juvenile recreation center, senior citizen recreation center.

CITIZENS: PREFERENCES, PARTICIPATION, AND COMMUNITY CONTEXTS

Citizens play critical, if varied, roles in numerous urban processes. Three types of research focus on policy preferences, participation, and political culture. Each links with research on national and state levels and is likely to continue for some time. There are, however, distinct advantages to analyzing these issues at the urban level and in the context of the larger project. There is considerably more empirical variation across localities and over time than at higher levels of aggregation. Such variation in local settings facilitates analysis of contextual characteristics. For example, a relatively upper-status suburb like Palo Alto differs from Newark in its political culture and rules of the game, its social problems and public services to deal with them, and in citizens' policy expectations and participation in local affairs. Even the same individual residing in Palo Alto or Newark is likely to shift his or her reactions more distinctly for local than for national elections and policy issues. Hence citizen activities and views concerning local matters are especially likely to be significantly related to the local context. If data are collected concerning important contextual characteristics with the rest of the project, they can then be linked to citizen reactions. There are also obvious cost savings in collecting data for these three aspects of citizens in a single survey.

At the theoretical level, at least two general and opposed traditions have informed empirical analysis of citizen preferences, participation, and political culture. One is the elite tradition from Schumpeter and Dahl, which stresses the critical role of elites as dynamic political actors and suggests that the average citizen exercises control more via support for a general political culture which he shares with elites—a sense of compromise, and so on. This approach stresses the importance of political culture, deemphasizes the role of specific policy preferences, and implies that political participation is likely to be low under any circumstance but tends to increase sporadically over specific issues. By contrast, the populist tradition building on the work of Anthony Downs and recent research on issue-voting suggests a far more active role for citizen policy preferences and tends to look for more consistent structure in such preferences. Also important here are relating these in turn to policy responsiveness of leaders, by analyzing leaders' policy preferences and actions and public service characteristics. A sort of intermediary position is one which emphasizes the characteristics of selected citizens who are especially informed and publically active. This blends in with the elite interpretation in Dahl (1961), or Converse (1964), or can be seen as offering potential for broader-based involvement of larger numbers of citizens (see Bennett, 1975). As a research strategy, this implies looking more carefully at the critical role of certain particularly active citizens, who may serve as "carriers" of beliefs which are important to explain the linkage of more general citizen preferences to public policy (see, on citizen-policy linkages, Hansen, 1978; Goodwin and Shepard, 1976; Luttbeg, 1974; Devine, 1970).

CITIZEN PREFERENCES

Since the late 1960s, citizen preferences have grown in salience, both for policy purposes and as a research topic. Urban programs like the Community Action Projects were initially important; similar concerns have continued and spread to more general programs, such as the

Community Development Block Grants and others. The "taxpayers' revolt" reflected in Proposition 13 and similar referenda further underscores how citizen preferences circumscribe government policies. Surveys of citizen policy preferences have become increasingly common—by political candidates, city agencies, and evaluation programs, as well as by social scientists.

Citizen preferences play an important role in many social-scientific theories, especially those of the public choice tradition often building on Anthony Downs (1957). Most public choice analyses remained theoretical until the mid-1970s, when efforts were made to study citizen preference data in conjunction with some of these citizen preference theories (for example, Page, 1977). Emprical efforts in turn have partly merged with a broad tradition of work on policy preferences.

Preference Structures. Labels differ in subfields, but terms such as values, preferences, wants, needs, utilities, and culture have largely overlapping referents (see, for general reviews, Clark, 1974; Converse, 1975; Georgescu-Roegen, 1968). The manner in which citizen preferences concerning public policies are or are not structured has been a central issue in several areas of research. Much classical decision theory dealing with isolated individuals assumed highly structured preferences (see Luce and Raiffa, 1957). Similarly stringent assumptions concerning citizen preferences were used in theoretical work by many economists. The essentially negative results of Arrow (1963) and Samuelson (1969) concerning the possibilities of defining a social welfare function have been widely accepted, given their assumptions (transitive, asymmetric, and complete preference orderings). These results no longer appear so nihilistic if something less than an ordering of "all states of the world" is sought; weaker assumptions can generate weaker orders or consistent social decisions (see Sen, 1970). Other aspects of preference structures that have been considered include unidimensionality, additivity, linearity, intensity, and related information level (for example, Fishburn, 1970).

A more empirical tradition of work on citizen preferences has also been concerned with their degree of structure or constraint, and thus how appropriately they may be used in a more general theory of the political process. In this regard, Converse's (1964) article has been especially important in its skepticism about the structure of policy preferences of the general citizenry. Although this view has been challenged using data from more recent years (see Pomper, 1972; Nie et al., 1976), the challengers in turn have been undercut by others who have shown that measurement procedures generated most of their "findings" (for example, Sullivan et al., 1978; Bishop et al., 1979). Another critique of the Converse position is that more unique patterns may exist for particular individuals or types of persons (Lane, 1962; Bennett, 1975). These arguments are recognized by Converse (1975: 83-94) as heightening the importance of studying policy preferences within the context of specific locales, where selected issues may be highly salient.

Comparing Citizen Preferences with Leaders and Public Policies. Besides the issues just posed, another complication is added when citizen preferences are compared to those of elected officials and to local public policies. The disparity between the policy desired by a citizen and that of the leader or actual policy has been analyzed as a loss function in the public choice tradition of Davis, Hinich, Kramer, and others (as in Riker and Ordeshook, 1973). Empirical work comparing leaders and citizens has found differences across issues, constituencies, and vulnerability of the leader to electoral defeat (see Miller and Stokes, 1963; Erikson, 1978; Kuklinski, 1978; Getter and Schumaker, 1978). Precisely what characteristics of leaders or policy should be compared with citizen preferences also is a subject of some discussion (Schumaker, 1979). A normative concern with representativeness often leads to a focus on comparing policy preferences of leaders with those of constituents (for example, Hansen, 1978) preferably studies using identical questions. Others have pointed out that a leader's personality, ethnicity, or social background may be more salient to many citizens than is his or her issue position (Dahl, 1961; Page, 1977). Still others have analyzed not the leader's stated preferences but actual policy choices, such as budget allocations or the delivery of services (see Schumaker and Getter, 1977; Ostrom et al., 1973).

Additional complications enter in studying citizen satisfaction with service delivery, such as including adequate data concerning citizen respondents, their neighbors, service providers, and elected officials. Hoffman and Clark (1979) and Rossi et al. (1976) have shown important effects of each such class of variables on citizen reactions to service delivery, which underscores the importance of collecting adequate data for each of the types of variables. This design distinguishes such studies from most others which have generally been restricted to one or a few urban areas.

In brief, a good deal of methodological experience has been accumulated, but we still do not have a single urban study that surveys citizens using a city "n" large enough to estimate with a reasonable standard error the preferences of citizens concerning urban public policy. However, studies of such limited scope do accumulate interesting results concerning measurement procedures that have been used to tap citizen preferences, such as Scott's (1976) Los Angeles survey comparing card sorts, ordinal rankings, self-ordering scales, and budget pies for assessments of public services; McIver and Ostrom's (1976) St. Louis study using budget pies and open-ended questions; Courant et al.'s (1979) Michigan study comparing various reactions to tax limitation initiatives; Gramlich's (1977) survey of formats concerning budget choices; and McKinney and MacRae's (1978) choice procedures extending some of these. Simpler question formats have been used in larger studies, such as preferences to "spend more, the same, or less" in each of several policy areas in 10 Urban Observatory cities in Fowler (1974); the NORC-GSS series for federal expenditures; and Schumaker et al. (1979) for leaders in 51 of the NORC-PCS cities. The simulation procedure developed in de Sola Pool et al. (1965) has been extended to states in Schneider (1976) and cities (Schumaker and Getter, 1977), by estimating local preferences using national surveys of citizens.

CITIZEN PARTICIPATION

While considerable effort has been invested in conceptualizing and measuring citizen preferences, in part due to their inherent complexity, participation seems simpler to study. Participation data have often been collected for national samples of individuals, or in individual cities, but seldom in a comparative urban framework.

The traditional form of participation is voting in elections, about which enormous quantities of data have been collected, or so it seems. It is a surprise to many to learn that only one study has ever been completed of a national sample of cities including voting turnout of citizens in local elections (Alford and Lee, 1968). And while it asked if the citizen voted or not, it did not ask party preference. Thus not a single study has been conducted to generate the classical voting data used to study national elections. Similarly, voting returns have never been assembled for American local elections by voting district or even entire cities, in the manner of Richard Scammon and others for national elections.

Nonvoting participation in voluntary associations and related activities were studied in a major national study by Verba and Nie (1972), which asked about both national and locally oriented activities.

Unfortunately, its sample size was too small to permit detailed city-level analysis.

Consequently, researchers have compromised in one of two general directions. Either they have used indirect measures of citizen participation for a large number of cities, or they have studied citizens directly in limited numbers of cities. For example, in an effort to test Coleman's (1955) hypotheses about participation and community conflict, Crain et al. (1969) used the legal form of government and whether or not a referendum was held as proxies for participation. Similarly, Lineberry and Fowler (1967), and others after them, have argued that leaders in cities with reform governments are more isolated from citizens; but despite methodologically important reanalysis of the relationship (as those by Wright, 1976, and Lyons, 1978), the basic hypothesis has not been tested with systematic data for citizen participation by form of government. Turk (1977) used the number of headquarters of voluntary associations in a city as a surrogate for the level of voluntary association activity in each city. Coles (1974) used an ACIR survey of leaders as informants about citizen participation patterns. Urban disturbance data have been coded from newspaper reports for estimates of numbers of participants, arrests, property damage, and so on (see Spilerman, 1971; Morgan and Clark, 1973). In each of these cases, direct data concerning citizen participation would have been valuable and could be used in future work along similar lines.

Following the second approach, direct studies of citizen participation have been completed in limited numbers of cities by Rossi et al. (1974), Coles (1974), Ostrom et al. (1973), Warren et al. (1974), Greenberg (1974), and others.

Partly as a consequence of these research designs, much of the empirical work to date has analyzed participation largely using characteristics of individual respondents. The "standard" model thus estimates participation rates as a function of respondents' income, education, and similar items. Further variables have been added and interactions among them considered in a few studies. But we have only fragmentary data by which to analyze urban structural characteristics as they relate to participation patterns. A large number of hypotheses, often supported by findings of a limited nature, are found in related literatures which add considerable richness to traditional participation studies. They point to variables that interact with participation rates, as well as to more diverse forms of input to the political process, and limitations on what governments do. Consider the following examples.

One fascinating finding in the unique Alford and Lee (1968) study was that it directly contradicted results of the "standard" model of individual citizen participation studies. That is, they found that the higher the mean educational level of citizens in a city, the lower the city's voting turnout. Although this has been considered simply an anomaly by some, it has also been interpreted as a consequence of greater homogeneity and civic quiesence in cities with more highly educated residents who are essentially satisfied with the status quo. Consistent with this pattern are a few case studies of individual cities were participation rates have reportedly reached 70 to 80 percent of the electorate—Gary or Cleveland at the time of racial transition to their first black mayor (Nelson and Meranto, 1977).

While this is suggestive of a pattern whereby urban structural characteristics (less homogeneous population, non-competitive political leadership, and so on) are likely to generate dissatisfaction of citizens (increased loss functions), and in turn higher participation rates, data have not been collected to analyze the specific linkages posited in this manner. This situation has meant that some of the most suggestive theories about political mobilization of citizens, and where and how their impacts are greatest, have remained either untested or examined using only rather poor data. For example, Coleman's early work on community conflict (1955), his (1971) volume on alternative strategies for mobilizing resources available to blacks, and his (1973) volume on a model of collective action all stress changes in patterns of resource activation necessary for individuals to affect the collectivity as a function of overall levels of participation and resource activation by others in the same system. Correspondingly, influence strategies shift and particular resources take on different values with such changes in system characteristics. Although these ideas have been conceptually important and examined in individual city contexts (for example, Marsden and Laumann, 1977), they have not been examined in the setting for which they were largely developed: comparing cities. Some not unrelated ideas of Dahl (1961) about "slack" and "pyramiding" of resources also remain largely untested, as do similar notions of Deutsch (1963) and Parsons (1969) about how the inflation or deflation of resources shifts with general political mobilization. Gamson's notions about resources, mobilization, and coalition building have been assessed primarily using experimental small groups (Gamson, 1968) and a creative but highly disparate selection of protest groups (Gamson, 1975). To analyze such matters for entire nation-states demands either a historical or cross-national study design; it is far more feasible to examine such ideas systematically using comparative urban data.

Originally developed to study cross-national politics, the concept of political culture was subsequently used to contrast styles of urban politics in typologies of Williams (1961), Banfield and Wilson (1963), Wilson and Banfield (1964, 1971), and Elazar (1970, 1972). The concept embraces citizen attitudes, beliefs, and other sentiments toward the political subsystem and one's own role as a political actor. It emphasizes normative aspects of the policy, rather than cognitive assessments or behavior. A recent review (Knoke, 1979) identified ten distinct dimensions of political values along which political subcultural variations might occur. However, the most fundamental dimension, as in the three above typologies, was one which ranged from "unitary" to "pluralistic."

In a unitary political subculture, politics is seen as a cooperative search for the interests of the entire community. By contrast, in a pluralistic (or "private-regarding") subculture, family needs and personal loyalties are paramount, and politics is just one of several avenues for the competitive advancement of personal or group interests. Government serves a more utilitarian peacekeeping function of balancing competing interests, with few obligations beyond preserving the marketplace for social and economic advancement.

While the national poltical culture is a historic blend of these ideal-typical subcultures, regions, communities, and individual citizens offer more distinct manifestations of each. The dominant political culture of a community influences its political subsystem by shaping adoption of formal governmental institutions, informal political participation, and constraining the range of appropriate activities of the local governments.

While initial formulations concerning urban political culture were based on fragmentary evidence, several studies have examined portions of these ideas, usually by restricting their focus to a particular set of variables reasonably comparable and available for numerous cities. Much work has thus analyzed the form of government, often classified in terms of "reform" characteristics such as nonpartisanship, at-large elections, and a city manager administration (Wolfinger and Field, 1966; Lineberry and Fowler, 1967; Lyons, 1978). This approach was often based on the assumption that such legal institutions reflected a reform "ethos," which could then be correlated with the socioeconomic characteristics of the population. However, Wilson and Banfield (1971) were among those most dissatisfied with this assumption; they advocated instead interviews with individual citizens, which led them to undertake in Boston probably the most thorough citizen survey to date of urban political culture. While some of the Boston items have been replicated in Detroit and Chicago (see Ewell and Meyers, 1978), data have never been collected to permit systematic analysis of city differences in political cultures using direct responses from individual citizens. Consequently, empirical studies of political culture have used quite indirect data to measure critical variables. Elazar (1966, 1972) completed an ambitious classification of the political culture of all U.S. states from a wide reading of general historical sources. Sharkansky (1969) correlated Elazar's measures with state voting turnout and government policy outputs. Other studies have sought to tap political culture using voting turnout (such as Alford and Lee, 1969: 154; Monroe, 1978). Schumaker and Getter (1977, 1978) have related the Elazar types, foreign stock, and other political culture measures to policy responsiveness of city governments using the NORC-PCS cities.

Many studies have linked political participation and attitude of citizens to ethnic and racial groups. These have ranged from national surveys of individuals (Greeley, 1974, for example) to intensive case study work on individual neighborhoods (see Suttles, 1968; Kornblum, 1974). In these and other studies, important differences in neighborhood characteristics related to participation of citizens as well as organized groups have been suggested but infrequently addressed with empirical data (Fischer, 1975; Warren, 1975).

Varied forms of participation have been studied, from voting and organized group membership to involvement in sit-ins, riots, rent strikes, boycotts of stores, and picketing. Most work on these processes, however, has tended toward explanations in terms of individual characteristics because of limitations of data to an individual city or national sample of citizens. Still, such studies have generated some interesting findings and hypotheses about group formation and collective behavior that demand comparative urban data to be adequately tested (see Genevie, 1978; Marx and Wood, 1975). For example, the incorporation of new political and other nontraditional forms of political involvement (for example, Gamson, 1975; Stinchcombe, 1975), but effects of such participation may be limited to cooptation of leaders and only symbolic victories if the goals of such groups challenge established procedures too seriously (Alford and Friedland, 1975). The considerable case study work on citizen participation since the 1960s, especially of the poor and blacks, has sometimes suggested minimal effects on public policy (see Peterson and Greenstone, 1973). But some recent studies do indicate significant changes (such as Karnig and Welch, forthcoming; Browning et al., 1978; Clark and Ferguson, forthcoming). Still, the comparative efforts are based on limited measures (percentage of black officials elected, changes in city expenditure patterns, etc.). The situation is basically the same as for citizen policy preferences: Numerous theories exist, many suggestive studies have been conducted in small numbers of cities, but little systematic comparative urban data exist to permit analysis of the theories where contextual variables play critical roles.

CANDIDATES FOR DATA COLLECTION

General social background characteristics (using standard questionnaire formats, such as those of the NORC-GSS and Michigan NES: age, sex, race, occupation, income, education, religion, national background, party affiliation, and so on)

Overall tax/service assessment (should the city government move toward higher service levels and taxes, or reductions; how large should changes be, if any? Several question formats could be compared for response patterns to this and related items below: continous scales, budget pies, card sorts, self anchoring scales, and new formats; see Scott, 1976)

Expenditure preferences by function (many formats are possible, but one widely used item asks "would you like your local government to spend more, the same, or less in each of the following service areas?" followed by a standard list of 15 or more service areas)

Service delivery preferences (constrained choices about mix of services citizens would prefer given different cost and service combinations, such as curbside garbage pick-up for X price per month versus alley pick-up for Y price, or different patrol frequencies for different prices)

Perceptions of service levels (what quantity and quality of different services do they see themselves now receiving)

Perceptions of service costs (what level of costs are perceived related to different services)

Evaluations of service levels (how satisfied are they with different services in their neighborhood, in other city neighborhoods, in other communities in metropolitan area; how satisfied are they with courtesy and honesty of service employees; are the benefilts greater than costs)

Recent contacts (whether victimized and follow-up questions on police contact, use of library, parks, recent road repair activities noticed, and the like)

Articulated demands (have they initiated contact with any of the six service agencies, with whom, about what, what result, have they contacted any political leader about any of the services, what happened)

Perception of access channels (if dissatisfied, who would they contact for different service problems, knowledge of access channels)

Mobility patterns (if recently moved into area, what factors taken into account in selection of neighborhood; if thinking about moving, what factors affect future moving plans)

Coproduction of service activities (participation in neighborhood watch program or other organized programs related to safety, efforts to recycle solid waste, investment in locks, lights, private security firms and devices, investment in private recreation equipment for back yard, participation in general neighborhood organization)

Participation in local organizations (political, religious, civic, unions, and so on; nature and extent of participation—contributing money, time, or other assistance)

Networks of personal contacts with major organizational and political leaders (adapt prodedures used in network studies of individual cities)

General norms and values (political culture, particularly concerning the scope of government, functions desired, and legitimate avenues for citizen and group input into public affairs; conception of social citizenship, appropriate role of private groups)

Assessments of major local groups (degree of support for such groups as racial minorities, the aged, municipal employees, government officials, local civic leaders)

Analogous items could be included for political leaders for major variables to permit comparison with citizen responses, especially of general social background, overall tax/service assessment, expenditure preferences by function, service delivery preferences, general norms and values, assessment of major local groups.

POPULATION REDISTRIBUTION AND THE LOCATION OF ECONOMIC ACTIVITIES

INTRODUCTION: GENERAL APPROACH AND TYPES OF DATA

In contrast to the situation for the three above sections, data collection is far less a problem for research on population and economic activities. The Census Bureau collects great quantities of data. But many census sources are complex to use, often involving elaborate programming in order to access and combine data elements from numerous tapes. Problems multiply rapidly for comparisons over time due to changes in boundaries, reporting prodedures, and other matters demanding quite detailed knowledge. The situation is especially discouraging for younger researchers. More experienced researchers have evolved solutions to many of these problems such that considerable benefits seem possible by having them help prepare date files to be available to all. As virtually no original data collection is involved, the costs of such activity are reasonably low. The 1980 census makes this an important time to plan for new analyses.

Discussions with researchers to date have led this section to take its present form. But again we stress that it is presented here as a basis for further discussion at the planning conferences.

Urban researchers concerned with population and economic activities have expressed interest in data for ten different types of analysis, as summarized in Figure 2.2. A first general point is that construction of hierarchical data files is desirable: including data for neighborhoods within cities, and cities within SMSAs, to permit analysis of contextual relationships across different levels. Neighborhoods, operationalized using census tracts, are the smallest units to be included. Cities are important as the major units of government which set tax rates, deliver services, and make most other public sector decisions affecting urban areas. However, private sector movements of population and jobs spill across such jurisdictional boundaries, making the metropolitan area (SMSA) an important unit. Two modes of analysis demand slightly different types of SMSA data. First is the intrametropolitan analysis, essentially of central cities of SMSAs compared to their suburban rings, involving such issues as deconcentration of population and jobs, suburban annexation, and commuting patterns. Second is intermetropolitan comparisons of entire SMSAs with one another, such as broad shifts like the rise of "Sunbelt" SMSAs. Finally, nonmetropolitan change patterns have taken on special interest since their relative growth in the 1970s has become clear; to help unravel causes of this growth we suggest comparison of metropolitan and nonmetropolitan areas.

For many population and economic base characteristics, data are readily available for almost all units in the United States, which suggests the economy of creating files involving the entire population of SMSAs and their suburban rings, and perhaps all cities over 25,000. A sample of

	Neighborhood	City	Intra-Metro	Inter-Metro	Non-Metro
Population	1	2	3	4	5
Economic Base	6	7	8	9	10

Figure 2.2 Ten Types of Foci for Research on Population and Economic Activities

200 census tracts might be appropriate. Although this comprises many data bits, many variables are identical or similar for the different levels (income, education, race and so on). Intermetropolitan analyses could use data for entire SMSAs. For intrametro purposes, the central cities and suburban rings of each SMSA would be listed separately. At the city level, the central cities of SMSAs and suburbs above tentatively 25,000 would be included, as well as certain independent cities. Neighborhoods would be selected from within the larger units. The nonmetropolitan data we plan to include are largely prepared already (by a University of Wisconsin project) and involve minimal additional work.

Some researchers concerned with jobs and population are likely to be interested in these data without those assembled in the rest of the project. Hence the creation of the larger files. From these larger files a subfile would be prepared for merging with the smaller number of units for which original data would be collected in the other parts of the project. Below are more details about specific data.

Data concerning population and economic activities are critical for many types of urban researchers. They are important for those concerned with political leadership, service delivery, and citizen characteristics, in that each of these processes is influenced by and in turn influences the types of persons and economic activities in an urban area. Hence the importance of including major population and economic data in many such analyses, even if sometimes only as controls.

Other researchers are interested in population and economic base characteristics in their own right. Although slightly less so than for the above three processes, much research on data on population and economic base questions has also used simple national totals or detailed investigation of one or a few urban areas. Hence the preparations of (the larger n) comparative urban data files can facilitate the work of these investigators, even if the files include only data from disparate census sources. Still, research in this area has very seldom used (noncensus) data concerning political leadership, service delivery, or citizen characteristics. Thus while losses of population or jobs have been documented as more frequent in certain urban areas, how much these have been influenced by such factors as poor public service delivery or unresponsive political leaders have largely remained subjects of speculation. The interpenetration of these various processes has generally been recognized at least in passing by researchers, but without adequate data they have not been empirically assessed. This would be possible with the (smaller n) files which merge data for all four processes.

RESEARCH QUESTIONS

Here follows a brief listing of some major concerns in these areas in which users have expressed interest. The next section considers problems of analyzing causes of these phenomena.

National population trends which affect urban areas and which are new or more pronounced in the 1970s than in earlier decades include a decline in the birth rate, fewer school-age children, and more older persons. These trends imply a general slowdown in growth, or absolute delcines in certain urban areas, even if all else were constant. One factor not constant, however, is size of the average household. This has declined, which raises demand for housing even while the total population in an area may be declining. Similarly, there has been substantial inmigration, particularly of Spanish-speaking persons, from abroad to many cities, although magnitudes are unclear, as many remain illegal aliens.

Metropolitan areas in the South and West have been growing while many in the northeast have been declining. The specifics of these patterns remain unclear, however, as they include the simultaneous movement of many types of persons and jobs, some in opposite directions. Nonmetropolitan areas have been growing, but farming activities have not. The larger SMSAs have generally been losing population, while many smaller ones are growing.

Within SMSAs, central cities are declining in population, income levels, and jobs compared to their suburbs for the nation as a whole. But within individual SMSAs the picture often differs, especially in smaller SMSAs.

Individual neighborhoods vary within cities, with much publicized "gentrification" of certain areas in such cities as Boston, New York, and Washington. Most discussions of such gentrification are based on observations of individual neighborhoods, however, and serious assessment of the extent of the phenomenon must await the 1980 census.

These various patterns in population and economic activity are bound up with local governments through the erosion of certain types of revenues, by generating pressures for increased or at least different public services, and often corresponding fiscal strain. Much media and policy discussion tends to tie these phenomena together, in part because social scientists have not documented clear causal linkages among them.

Indeed, most of this section has simply listed a number of phenomena that reflect some major results of census reports primarily for the 1970s (see Alonso, 1978; Beale, 1975; Berry and Dahlmann, 1977; Tucker, 1976; U.S. Bureau of the Census, 1975a, 1975b). Most intercensal results, however, are drawn from Current Population Surveys, which are based on a sample of the total U.S. population. Its sample size is too small to include more than total population changes for most individual urban area, such that many results are published only by region or by population size of SMSAs. Hence evidence with which to analyze more carefully the causal structure of certain of these phenomena must await the 1980 census.

CAUSAL ANALYSES OF POPULATION AND ECONOMIC ACTIVITIES

While the last section offered an overview in a more descriptive mode, most of these phenomena have also been studied for many years from a more explicit causal modeling perspective. The availability of critical data and the complexities of such work usually mean that careful analyses lag several years behind the descriptive reports of new developments. Hence results from earlier analyses of similar trends, or situations observed using noncensus data in individual metropolitan areas, are often used to interpret initial descriptive results of new trends. This is normal in all scientific fields, of course, but has led to particularly challenging questions being raised in recent years about some widely accepted past results. One of the most dramatic is the nature of relationships between population and job changes.

Most standard theories of migration have explained population movement by location of firms (from Lösch through Alonso, 1968; Muth, 1969; and Kain, 1970; 1975.) Specifically, rural-urban migration in most of the world—and in American cities from the nineteenth century to at least the 1950s—has generally been explained by the growth of jobs in urban areas and the declining demand for farm labor due to technological improvements in agriculture (Shaw, 1975). Similarly, the rise of suburbs and decline of central cities have been explained by communication changes like the telephone, airplanes, and interstate highways, which reduced the importance of personal contacts; by the lower costs of single-story assembly and warehouse areas; and by feasibility of access provided by rail-truck piggyback combinations. These technological developments encouraged the location of manufacturing in suburban and other low-density areas; people followed their jobs to these areas (see Meyer et al., 1965; Berry and Kasarda, 1977; Vaughan and Vogel, 1979).

With the emerging documentation of the Sunbelt, much public discussion has projected these past trends, encouraging tax advantages by local governments to large firms, and focusing media attention on such phenomena as movement of corporate headquarters from New York to Houston—implying that relocation of major corporations was in turn important in generating shifts in population. This interpretation has been challenged in three fundamental ways by recent research. One is more careful modeling of the process of urban job creation and loss, which has shown remarkably high rates of turnover. That is, large numbers of jobs disappear with firms that die each year, and are replaced by new ones. Further, most changes in urban jobs are explained not so much by the death rate (or outmigration) of firms as by the differential birth rate of new firms in growing areas. And most of these involve small and medium-sized firms, such that national corporations that shift their headquarters may affect relatively small portions of the overall movement toward the South and West (see Birch et al., 1977; Vaughan and Vogel, 1979).

The second example of recent work that has helped redefine the relationship between jobs and population is simultaneous equation modeling of interregional movements (for example, Muth, 1971; Greenwood, 1975). These studies have found substantial simultaneous impact of jobs on population and population on jobs; the striking result is that the latter effects may be as strong or stronger. This implies that persons often move to areas like San Diego and Phoenix, and, once there, jobs are created by or for them, rather than vice versa, as has usually been assumed.

The third example concerns the increasing role of public policy in affecting both population and economic location decisions. Such public policies have grown considerably more numerous and diverse in recent years, such that earlier research on these issues may be out of date. Increases in local and state tax rates, changing intergovernmental fiscal flows, massive manpower and urban development programs (such as CETA, CDBG, UDAG, ARFA, and others) have been documented as having substantial effects in certain individual cities, as well as in efforts to review the national picture (Vernez et al., 1979; National Commission for Manpower Policy, 1978; Tolley et al., 1979). But the research on these issues to date has typically concluded that data remain inadequate for careful analysis. Here, in particular, the several sections of this proposal complement one another, in that data on fiscal policy, service delivery, and other public sector variables can be combined to assess their effects on location decisions by private individuals and firms.

Still, these new research findings have been greeted with less than total enthusiasm by specialists in each area, such that we can expect to see a good deal of effort involved in their reassessment. If these new results are at least partly sustained, however, they imply respecifying many existing models and shifts in the types of variables likely to play major causal roles. For example, for many residential choice models the role of such factors as climate, proximity of oceans, beaches, or mountains, and tax burdens and public services of local government may take on heightened importance, while economic activity characteristics may decrease in their importance. (See Bonczar, 1979, for evidence indicating that this is occurring.)

Many other interesting developments are emerging in analyzing population and economic base phenomena. A number of phenomena which have been considered relatively well documented have been clarified in recent work. For example, the relative growth of suburbs compared to their central cities has been shown to have emerged about 1900 rather than in the interwar years, as was more generally accepted. The earlier date derives from careful compilation of data concerning annexation in the earlier years (Kasarda and Redfern, 1975). Suburban development of jobs combined with continuing residence of many poor and blacks in central cities leads to an ecological mismatch of substantial social and economic cost (see Gold, 1972; Harrison, 1974; Zimmer, 1975). There seems to be considerable vitality in certain types of functions in older cities, however, such as in mass coommunications, accounting, law, and certain others (Hoskin, 1973). The role of public service provision in such continued vitality of certain older neighborhoods is beginning to be addressed (for example, Friedland and Bielby, 1979). The degree to which suburbs exploit central cities by virtue of increasing costs more than they generate in revenue is a topic of increasingly detailed analysis (see Kasarda, 1972; Neenan, 1970). Posing these questions in a broader perspective to include political impacts and relations among different sorts of suburbs as well as metropolitan areas further extends these issues (Logan, 1979; Schneider, 1979; Moloch, 1976). Research tools ranging from classical regression to simultaneous equations to simulations to new procedures have been emerging (Greenwood, 1975; Sweet, 1972; Rossi, 1979; Wright et al., 1979). The obvious policy impacts of work in this area have led to research procedures which include explicit attention to policy impacts of major variables such as environmental controls, zoning tax burdens, and social service benefits (Tolley et al., 1979; Katzman, 1978).

DATA SOURCES

An excellent site for initial data file development of demographic and economic base indicators is the Institute for Research in Social Sciences at the University of North Carolina. Through a series of substantial urban data consortium grants from the National Science Foundation, U.S. Office of Education, the U.S. Bureau of the Census, IRSS has developed the largest machine-readable urban demographic data library of any university in the nation. The bulk of the economic place data described above are also on computer retrieval files at IRSS, as well as the full inventory of ICPSR urban data sets, HUD surveys, and special urban data libraries purchased from the University of Wisconsin, Madison, and the University of California at Berkeley. John D. Kasarda has agreed to supervise file development of demographic and economic place data if IRSS is selected as the site for initial file development for this portion of the project.

Basic demographic changes occurring for SMSAs, central cities, and suburban rings (for example, population growth, changing density patterns, changing racial, occupation, and income composition) could be documented through creation of files derived from the 1940, 1950, 1960, 1970, and forthcoming 1980 *Censuses of Population* and numerous *Current Population Surveys.*

Data for analysis of the changing distribution of retail, wholesale, and business service activities (including number of establishments, employees, and sales or receipts, by nature of business) could be obtained from the 1939, 1948, 1954, 1958, 1963, 1967, 1972, and 1977 *Censuses of Business,* supplemented by information provided in County Business Patterns, 1946 to 1979. The suburbanization of manufacturing activity (including number of establishments, production and other manufacturing employees, and value added) could be assessed with data obtained from the 1939, 1947, 1954, 1963, 1967, 1972, and 1977 *Censuses of Manufacturers* and the *Annual Survey of Manufacturers,* 1958 to 1979.

Data relating to the centripetal buildup in the central cities of governmental, managerial, professional, clerical, and other white-collar functions could be obtained from published and unpublished data contained in the place of work reports from the 1960, 1970, and 1980 *Censuses of Population.* These could be supplemented with data on the growth of administrative and central office functions in central cities and suburban rings from associations like the Building Owners and Managers Association, International (Chicago, Illinois).

Commuting between central cities and suburban rings (including reverse commuting from city to suburb) by place of residence, race, income, occupation, and industry of commuters could be examined by merging published and unpublished data on journey to work derived from the 1960, 1970, and 1980 *Censuses of Population.* More detailed social, occupational, and demographic data on commuters by broad categories of residence and place of work could be obtained from the 1960, 1970, and 1980 compatible 1/100 Public Use Samples and the 1975-76 Department of Transportation Supplementary survey on travel to work.

Relative changes in city revenue by source (such as property taxes, sales taxes, and intergovernmental grants) and public expenditures by function (for example, welfare payments and expenditures for police, fire, highway, sanitation, recreation, and administration) could be documented with public finance data obtained from the 1957, 1962, 1967, 1972, and 1977 *Censuses of Governments* and annual compendia of *City Government Finances,* 1947 to 1979.

Data on employment status by race, education, occupation, and industry of employment (or last job held by unemployed) could be obtained from publications of the 1950, 1960, 1970, and 1980 *Censuses of Population* and from periodic reports of the Bureau of Labor Statistics. The 1970 and 1980 *Census of Population* fourth-count summary tapes would be particularly useful in locating high unemployment sectors and poverty pockets within each metropolitan area and examining their ecological characteristics. In addition, the 1960, 1970, and 1980 Public Use Samples provide data for individual's employment status by place of residence, place of work (central city-suburban ring dichotomy), occupation, industry of employment, and a broad variety of other useful variables such as race, income, and educational background.

The metropolitan communities for which the majority of data would be gathered are all SMSAs currently delimited by the Bureau of the Census (N=270). Data would be obtained for central-city components and each metropolitan county. To improve longitudinal comparability, SMSA and central-city boundaries would be adjusted to constant criteria (fixed boundaries over time) according to the latest definitions. Since SMSAs are comprised of counties (except in New England where towns and cities are used) adjustments would be made by adding or deleting appropriate county data at a particular date. The ecological structure of SMSAs created since 1950 would be reconstructed historically on the basis of their county and place data for prior decades. We would reconstruct SMSA boundaries for New England in terms of counties to improve comparability with SMSAs in other regions of the nation.

The data should be adjusted for central cities and suburban rings to account for intrametrapolitan boundary changes over time. For example, annexation and consolidation systematically understate suburban ring growth and overstate central city growth. Methodological procedures for taking annexation into account are presented in Kasarda (1976). When demographic and economic data do not correspond temporally (for example, manufacturing data for 1963 or 1972 and population data for 1960 or 1970), standard interpolation and extrapolation can be carried out to adjust the data to comparable dates. Moreover, in longitudinal analyses of economic activity utilizing monetary indicators (such as sales, receipts, value added, public service expenditures), the indicators should often be converted to constant dollars using appropriate price deflators with the price deflators included in the file.

We also plan to construct system files for cities and neighborhoods (census tracts). The 1980 Census of Population and Housing will, for the first time, provide a substantial amount of demographic and housing data for local neighborhood units. We would like to explore the feasibility of incorporating hierarchically much of these 1980 neighborhood data in the city records. We would explore the possibility of using fourth-count tract data from the 1970 census to reconstruct longitudinally comparable 1970 neighborhood data.

For cities over a certain population size, perhaps we could include virtually the complete range of demographic and economic variable described above for central cities and metropolitan counties. While some economic data will not be available for all these places (because of Census Bureau disclosure policy), most data are possible to obtain.

Social and economic data similar to those for the larger units will be included for census tracts, although such data are generally less available, due to concern for anonymity and because certain phenomena either do not exist at the tract level or are similar for the entire city (such as city government functional responsibility).

SOCIOPOLITICAL CONTEXT DATA

This heading does not indicate a separate process, but a category of variables that involves many of the above four processes. The sections above have alluded repeatedly to specific characteristics of the sociopolitical context as appropriate for each process. However, in turning from theories and past research to data collection strategies, it seems useful to begin to list variables concerning the sociopolitical context of cities that may be collected from local institutional sources, rather than from political leaders, agency officials, or citizens. These data in many instances are fairly readily available and relatively inexpensive to collect. Still, as a general strategy, we plan to consider collecting only those data elements which relate to some hypothesis about one of the basic processes discussed above; otherwise the task could become too undefined. Even more than the above lists of variables, the items below are intended to be only suggestive. Possible variables:

Voting patterns by election district (these could be completed in the same general manner as the Scammon *America Votes* series, to include turnout, votes for each candidate, and party affiliation—for partisan elections)

Legal structure of the city government (city manager, at-large elections, nonpartisanship, powers of and council, civil service, etc. These have traditionally been drawn from surveys by the International City Management Association, but some recent work has questioned the adequacy of some standard ways of reporting and analyzing these data; for example, Robinson and Dye, 1978.)

Functional responsibilities of major governments in the SMSA (for each function or service area, ascertain whether provided and/or produced by city, special district, overlapping government, contract with other government, contract with private agency, and whether any enclave producers or service within city)

Voluntary association activities (for a list of voluntary associations: presence in the city, staff, budget, membership—if appropriate, social characteristics of members, such as income or race)

Social movement activities (including fluoridation controversies in the 1950s, civil rights demonstrations, and urban disorders in the 1960s, ecological movements activities, municipal employee strikes, homeowner, renter or taxpayer's related activities—rent strikes, refusal to pay property taxes, and the like)

Mass media (presence of various types of media in the metropolitan area, amount and type of coverage of local news, affiliations with national networks, other chain memberships, ownership including cross-media ownership)

UNITS OF ANALYSIS, SAMPLES, DATA STRATEGIES

While the specific research design of the project must await further discussion in the conferences, discussions to date have led to certain possibilities consistent with analysis of the above processes. These suggest an integrated research design combining data from several types of sources in a manner permitting multiple analysis strategies.

(1) Inventory Existing Data Sets. Certain urban data sets are already well known and documented; others, far less so. A reasonably inexpensive activity is to inventory systematically the major urban data sets whose quality and scope are sufficient to make them of continuing importance. Many such data sets are currently stored and distributed via the Inter-University Consortium for Political and Social Research in Ann Arbor; we plan to work with ICPSR to help inventory and disseminate urban data more broadly.

One report of urban data sets available through ICPSR was prepared in 1975; after new studies are added, this could be updated and distributed to urban researchers. More detailed documentation concerning specific variables in the studies could be included than in the past. Contributions from a large number of scholars in different disciplines should be solicited to help locate important studies or data sources which should be inventoried and to evaluate the validity and reliability of available data.

In certain instances data from such earlier studies can be merged with new data. In other instances this may not be possible, but it is still useful to conduct at least preliminary analysis of similar items from an earlier survey in designing new research.

The types of data discussed below can be considered elements of a multilevel research design. Some Chicago conference participants expressed interest in analyzing SMSAs, cities, and neighborhoods in relationship to each other. For example, metropolitan versus individual city rates can be studied in terms of population growth. Although this implies a complex research design, it is less expensive than one might imagine, as many data are available from the Census Bureau.

(2) Prepare Integrated Files of Census Data. The census remains the leading source of especially population and economic activity data. Specific data problems are listed below, but here we note simply that in constructing data files from census tapes it is often sensible to work with the universe of available urban data. For example, it seems reasonable to create files of all 200 plus SMSAs, and all 800 plus cities over 25,000 (specific n's vary by year), as many census data for these units are readily available. For some purposes it may also be worthwhile to include data from cities smaller in population size. Further details are introduced below concerning the integrated Census data file.

(3) Collect Original Data from a Hierarchical Sample of SMSAs, Cities, Neighborhoods, and Citizens. For purposes of addressing questions raised by different research traditions, it is important to include certain data from four different levels: metropolitan areas (SMSAs), cities, neighborhoods, and individual citizens. At the Chicago conference and in subsequent discussions emerging from it, several possible approaches have been considered for constructing an appropriate sample that would include all four levels. Although working memoranda have been developed concerning this matter in detail, at this point it seems appropriate simply to indicate that it might include some 20 SMSAs, 50 to 100 cities, and 50 to 100 neighborhoods, hierarchically selected in such a way that the lower-level units would fall inside one of the higher-level units in the sample. The Permanent Community Sample of the National Opinion Research Center, consisting of subsamples of the NORC Primary Sampling Units, has been studied repeatedly since the 1960s. Various PCS studies have included SMSAs, cities (central cities and suburbs), neighborhoods, and individual citizens. The NORC-PCS thus provides useful baseline data and a possible sampling frame for future work.

While important data for each level can come from Census Bureau sources, original data collected from published documents, interviews (personal or telephone), observation, and other procedures would also be desirable.

(4) Sample Perhaps Four Cities for Detailed Analysis. Certain measures such as networks of organizational contacts, the evolution of public issues, and other data from local records could be productively studied in detail in a limited number of cities. If at least abbreviated versions of similar indicators are collected in the sample of 50 to 100, then the representativeness of the four can be suggested. Criteria for sampling cities for intensive analysis were discussed by several conference participants. Having a small number of cities for exploratory work on questionnaire design and for some of the more expensive multiple indicator measures also makes methodological sense in the context of a national sample.

REFERENCES AND SELECTED BIBLIOGRAPHY

AHLBRANDT, R. S. (1973) Municipal Fire Protection Services: Comparisons of Alternative Organizational Forms. Beverly Hills, CA: Sage.

ALFORD, R. R. and R. FRIEDLAND (1975) "Political participation and public policy" pp. 429-479 in Annual Review of Sociology, Vol. 1. Palo Alto, CA: Annual Reviews.

ALFORD, R. R. and E. C. LEE (1968) "Voting turnout in American cities." American Political Science Review 62 (September): 796-813.

ALONSO, W. (1978) "The current halt in the metropolitan phenomena," pp. 23-41 in C. L. Leven (ed.) The Mature Metropolis. Lexington, MA: D. C. Heath.

———(1964) Location and Land Use. Cambridge, MA: Harvard University Press.

ANTON, T. (1979) "Data systems for urban fiscal policy." Presented at the Conference on Comparative Urban Research, University of Chicago.

ARROW, K. J. (1963) Social Choice and Individual Values. New York: John Wiley.

BAHL, R. W. (1968) "Studies on determinants of expenditures: a review," pp. 184-207 in S. Mushkin and J. Cotten (eds.) Functional Federalism. Washington, DC: George Washington University Press.

BANFIELD, E. C. and J. Q. WILSON (1963) City Politics. Cambridge, MA: Harvard University Press/MIT Press.

BEALE, C. L. (1975) The Revival of Population Growth in Nonmetropolitan America. Report #605, Economic Research Service, U.S. Department of Agriculture.

BENNETT, W. L. (1975) The Political Mind and the Political Environment. Lexington, MA: D. C. Heath.

BERRY, B.J.L. and D. C. DAHLMANN (1977) Population Redistribution in the United States in the 1970s. Washington, DC: Assembly of Behavioral and Social Sciences, National Research Council, National Academy of Sciences.

BERRY B.J.L. and J. D. KASARDA (1977) Contemporary Urban Ecology. New York: Macmillan.

BIDERMAN, A. D. (1966) "Social indicators and goals," pp. 68-153 in R. Bauer (ed.) Social Indicators. Cambridge, MA: MIT Press.

BIRCH, D. L. et al. (1977) The Behavioral Foundations of Neighborhood Change. Cambridge: Joint Center for Urban Studies of Harvard and MIT.

BISHOP, G. F. et al. (1979) "Questions about question wording." American Journal of Political Science, 23 (February): 187-192.

BLAIR, L. H. and A. L. SCHWARTZ (1972) How Clean Is Our City? Washington, DC: The Urban Institute.

BONCZAR, T. R. (1979) Changing Migration Patterns. Ph.D. thesis, University of Chicago. (unpublished)

BOOTH, A., D. R. JOHNSON, and H. M. CHOLDIN (1977) "Correlates of city crime rates: victimization surveys versus official statistics." Social Problems 25 (December): 187-197.

BOOTS, A. et al. (1972) Inequality in Local Government Services. Washington, DC: The Urban Institute.

BORCHERDING, T. E. [ed.] (1977) Budgets and Bureaucrats: The Sources of Government Growth. Durham, NC: Duke University Press.

BRIEGER, R. L. (1979) "Toward an operational model of community elite structures." Quality and Quantity 13: 21-57.

BROWNING, R. P., D. R. MARSHALL, and D.H. TABB (1978) "Responsiveness to minorities: a theory of political change in cities." Presented at the American Political Science Association annual meeting.

BURT, R. S. (1979) "Spatial models of community leadership" and "Comparative power structures in American communities." Presented at the Conference on Comparative Urban Policy Research, University of Chicago.

——— (1977) "Power in a social topology," p. 251-354 in R. J. Liebert and A. W. Imerschein (eds.) Power, Paradigms, and Community Research. Beverly Hills, CA: Sage.

CAPUTO, D. A. and R. L. COLE (1977) "City officials and mailed questionnaires." Political Methodology 4: 271-287.

CHAPMAN, J, W. Z. HIRSCH, and S. S. SONENBLUM (1975) "Crime prevention, the police production functions, and budgeting." Public Finance 30: 197-215.

CLARK, T. N. [ed.] (1976) Citizen Preferences and Urban Public Policy: Models, Measures, Uses. Beverly Hills, CA: Sage.

——— (1975a) "Cities differ—but how and why?: Inputs to national urban policy from research on decision-making in 51 American municipalities." Report to Office of Policy Development and Research, U.S. Department of Housing and Urban Development, September.

——— (1975b) "The Irish ethic and the spirit of patronage." Ethnicity 2: 305-359

——— (1974) "Can you cut a budget pie?" Policy and Politics 3 (December): 3-32.

——— [ed.] (1968) Community Structure and Decision-Making. New York: Thomas Y. Crowell.

——— and L. FERGUSON (forthcoming) Political Processes and Urban Fiscal Strain.

COLES, R. L. (1974) Citizen Participation and the Urban Policy Process. Lexington, MA: D. C. Heath.

COLEMAN, J. S. (1973) The Mathematics of Collective Action. Chicago: AVC.

——— (1971) Resources for Social Change. New York: Wiley-Interscience.

——— (1955) Community Conflict. New York: Free Press.

Committee for Economic Development (1970) Reshaping Government in Metropolitan Areas. New York: Committee for Economic Development, Research and Policy Committee.

CONVERSE, P. E. (1975) "Public opinion and voting behavior" pp. 75-170 in F. I. Greenstein and N. W. Polsby (eds.) Handbook of Political Science, Vol. 4. Reading, MA: Addison-Wesley.

COURANT, P. N., E. M. GRAMLICH, AND D. L. RUBINFELD (1979) The Tax Limitation Movement. Discussion Paper 141. Ann Arbor, MI: Institute of Public Policy Studies.

CRAIN, R. L., E. KATZ, AND D. B. ROSENTHAL (1969) The Politics of Community Conflict. Indianapolis: Bobbs-Merrill.

CRECINE, J. P. (1969) Governmental Problem-Solving: A Computer Simulation of Municipal Budgeting. Chicago: Rand McNally.

DAHL, R. A. (1961) Who Governs? New Haven, CT: Yale University Press.

de SOLA POOL, I., R. P. ABLESON, and S. POPKIN (1965) Candidates, Issues, and Strategies. Cambridge: MIT Press.

DEUTSCH, K. W. (1963) The Nerves of Government. New York: Free Press.

DEVINE, D. J. (1970) The Attentive Public. Chicago: Rand McNally.

DOWNS, A. (1957) An Economic Theory of Democracy. New York: Harper & Row.

ELAZAR, D. J. (1972) American Federalism: A view from the States (2nd. ed.) New York: Thomas Y. Crowell.

——— (1970) Cities of the Prairie. New York: Basic Books.

——— (1966) American Federalism: A View from the States. New York: Thomas Y. Crowell.

ERIE, S. P., J. J. KIRLIN, and F. F. RABINOVITZ (1972) "Can something be done? Propositions on the performance of metropolitan institutions," pp. 7-41 in L. Wingo (ed.) Reform of Metropolitan Governments. The Governance of Metropolitan Regions, Vol. 1. Washington DC: Resources for the Future.

ERIKSON, R. S. (1978) "Constituency opinion and congressional behavior: a reexamination of the Miller-Stokes representation data." American Journal of Political Science 22 (August): 511-535.

EULAU, H. and K. PREWITT (1973) Labyrinths of Democracy: Adaptations, Linkages, Representation and Policies in Urban Politics. Indianapolis: Bobbs-Merrill.

EWELL, P. and L. C. MEYERS (1978) "Some notes on political culture in Chicago." Draft paper, December.

FAURI, D. P. (1978) "Public service as a service to clients." American Behavioral Scientist 21 (July/August): 859-880.

FISCHER, C. S. (1975) The Urban Experience. New York: Harcourt Brace Jovanovich.

FISHBURN, P. C. (1970) Utility Theory for Decision-Making. New York: John Wiley.

FOWLER, F. J., Jr. (1974) Citizen Attitudes Toward Local Government, Services and Taxes. Cambridge, MA: Ballinger.

FRIEDLAND, R. and W. T. BIELBY (1979) "The power of business in the city." Presented at the Conference on Comparative Urban Policy Research, Chicago, April.

FRIEDMAN, L. B. (1975) Budgeting Municipal Expenditures. New York: Praeger.

FRIESEMA, H. P. (1966) "The metropolis and the maze of local government." Urban Affairs Quarterly 2 (December): 68-90.

FROHLICH, N., J. A. OPPENHEIMER, and O. R. YOUNG (1971) Political Leadership and Collective Goods. Princeton, NJ: Princeton University Press.

FUCHS, V. R. (1973) The Service Economy. New York: Basic Books.

GALASKIEWICZ, J. (1979) Exchange Networks and Community Politics. Beverly Hills, CA: Sage.

GAMSON, W. (1975) The Strategy of Social Protest. Homewood, IL: Dorsey Press.

——— (1968) Power and Discontent. Homewood, IL: Dorsey Press.

——— (1964) "Experimental studies in coalition formation," L. Berkowitz, (ed.) in Advances in Experimental Social Psychology. New York: Academic Press.

GARN, H. A. et al. (1976) Models for Indicator Development: A Framework for Policy Analyses. Washington, DC: The Urban Institute.

GENEVIE, L. E. [ed.] (1978) Collective Behavior and Social Movements. Itasca, IL: F. E. Peacock.

GEORGESCU-ROEGEN, N. (1968) "Utility," pp. 236-267 in International Encyclopedia of the Social Sciences, Vol. 16. New York: Free Press/Macmillan.

GETTER, R. W. and P. D. SCHUMAKER (1978) "Contextual bases of responsiveness to citizen preferences and group demands." Policy and Politics 6 (March): 249-278.

GOLD, N. N. (1972) "Mismatch of jobs and low-income people in metropolitan areas and its implications for the central-city poor," in S.M.Mazie (ed.) U.S. Commission on Population Growth and the American Future, Population Distribution and Policy, Vol. V of Commission Research Reports. Washington, DC: Government Printing Office.

GOODWIN, R. K. and W. B. SHEPARD (1976) "Political processes and public expenditures." American Political Science Review 70 (December): 1186-1198.

GRAMLICH, F. W. (1977) "The demand for clean water." National Tax Journal 30 (June): 183-195.

GREELEY, A. M. (1974) Ethnicity in the United States. New York: Wiley-Interscience.

GREENBERG, S. B. (1974) Politics and Poverty. New York: Wiley-Interscience.

GREENSTONE, J. D. and P. E. PETERSON (1973) Race and Authority in Urban Politics. New York: Russell Sage.

GREENWOOD, M. J. (1975) "A simultaneous-equations model of urban growth and migration." Journal of the American Statistical Association 70 (December): 797-810.

GREENWOOD, P. W. and J. PETERSELIA (1977) Criminal Careers of Habitual Felons. Santa Monica, CA: Rand Corporation.

GRIMES, M. D., C. M. BONJEAN, J. L. LYON, and R. L. LINEBERRY (1976) "Community structure and leadership arrangements: a multidimensional analysis." American Sociological Review 41 (August): 706-725.

GROENNINGS, S. et al. [eds.] (1970) The Study of Coalition Behavior. New York: Holt, Rinehart & Winston.

HANSEN, S. B. (1978) "Linkage models, issues, and community politics." American Politics Quarterly 6 (January): 3-28.

HARRISON, B. (1974) Urban Economic Development. Washington, DC: The Urban Institute.

HARVEY, D. (1973) Social Justice and the City. Baltimore, MD: John Hopkins University Press.

HAWLEY, A. H. (1957) "Metropolitan population and municipal government expenditures in central cities," pp. 773-782 in P. K. Hatt and A. J. Reiss, Jr. (eds.) Cities and Society. New York: Free Press.

HINCKLEY, B. (1979) "Twenty-one variables beyond the size of winning coalitions." Journal of Politics 41 (February): 192-213.

HIRSCHMAN, A. A. (1970) Exit, Voice and Loyalty: Responses to Declines in Firms, Organization, and States. Cambridge, MA: Harvard University Press.

HOFFMAN, W. L. and T. N. CLARK (1979) "Citizens' preferences and urban policy types," pp. 85-106 in J. P. Blair and D. Nachmias (eds.) Fiscal Retrenchment and Urban Policy. Beverly Hills, CA: Sage.

HOSKIN, F. P. (1973) The Functions of Cities. Cambridge, MA: Schenkman.

HUNTER, F. (1953) Community Power Structure. Chapel Hill: University of North Carolina Press.

JACOB, P. et al. (1971) Values and the Active Community. New York: Free Press.

JONES, B. et al. (forthcoming) Detroit Services Delivery.

KAIN, J. F. (1975) Essays on Urban Spatial Structure. Cambridge, MA: Ballinger.

——— (1970) "The distribution and movement of jobs and industry," pp. 1-44 in J. Q. Wilson (ed.) The Metropolitan Enigma. Garden City, NY: Doubleday.

Kansas City Police Department (1977) Response Time Analysis. Kansas City, MO: Board of Police Commissioners.

KARNIG, A. and S. WELCH (forthcoming) Book of Black Leadership. Chicago: University of Chicago Press.

KASARDA, J.D. (1976) "The changing occupational structure of the American metropolis: apropos the urban problem," pp. 113-135 in B. Schwartz (ed.) The Changing Face of the Suburbs. Chicago: University of Chicago Press.

——— (1972) "The impact of suburban population growth on central city functions." American Journal of Sociology 77 (6): 1111-1124.

——— and G. REDFERN (1975) "Differential patterns of urban and suburban growth in the United States." Journal of Urban History 2 (November): 43-66.

KATZMAN, M. T. (1978) The Quality of Municipal Services, Central City Decline, and Middle-Class Flight. Research Report R78-1. Cambridge: Department of City and Regional Planning, Harvard University.

KNOKE, D. (1979) "Stratification and the dimensions of American political orientations." American Journal of Political Science 23 (November).

KORNBLUM, W. (1974) Blue Collar Community. Chicago: University of Chicago Press.

KOTTER, J. P. and P. R. LAWRENCE (1974) Mayors in Action. New York: Wiley-Interscience.

KRAEMER, K. and W. DUTTON (1973) "Technology innovation and local government." Policy and Politics 6 (September): 27-49.

KUKLINSKI, J. H. (1978) "Representations and elections: a policy analysis." American Political Science Review 72 (March): 165-177.

LANE, R. (1962) Political Ideology. New York: Free Press.

LAUMANN, E. O. and F. U. PAPPI (1976) Networks of Collective Action. New York: Academic Press.

LEVY, F., A. MELTSNER, and A. WILDAVSKY (1974) Urban Outcomes: Schools, Streets, Libraries. Berkeley: University of California Press.

LIEBERT, R. J. (1976) Disintegration and Political Action: The Changing Functions of City Government in America. New York: Academic Press.

——— (1974) "Municipal functions, structure, and expenditures: reanalysis of recent research." Social Science Quarterly 54 (March): 765-783.

——— and A. W. IMERSCHEIN [eds.] (1977) Power, Paradigms, and Community Research. Beverly Hills, CA: Sage.

LINEBERRY, R. L. (1977) Equality and Urban Policy: The Distribution of Municipal Public Services. Beverly Hills, CA: Sage.

———and E. P. FOWLER (1967) "Reformism and the public policies in American cities." American Political Science Review 61 (September): 701-716.

LOGAN, J. R. (1979) "Metropolitan political economy." Presented at the Conference on Comparative Urban Policy Research, Chicago, April.

LUCE, R. D. and H. RAIFA (1957) Games and Decisions. New York: John Wiley.

LUTTBEG, N. B. [ed.] (1974) Public Opinion and Public Policy. Homewood, IL: Dorsey Press.

LYONS, W. (1978) "Reform and response in American cities: structure and policy reconsidered." Social Science Quarterly 59 (June): 118-132.

MacGILLVRAY, L. et al. (1977) Final Report to the National Science Foundation on Fire Services Study, Phase I. Durham, NC: Research Triangle Institute.

McIVER, J. P. and E. OSTROM (1976) "Using budget pies to reveal preferences: validation of a survey instrument," pp. 87-110 in T. N. Clark [ed.] Citizen Preferences and Urban Public Policy. Beverly Hills, CA: Sage.

McKINNEY, M. W. and D. MacRAE, Jr. (1978) "Survey assessments of consumer demands for publicly supplied goods: recreational facilities." Chapel Hill: University of North Carolina. (mimeo)

MARSDEN, P. and E. O. LAUMANN (1977) "Collective action in a community elite," pp. 199-250 in R. J. Liebert and A. W. Imershein (eds.) Power, Paradigms, and Community Research. Beverly Hills, CA: Sage.

MARX, G. T. and J. L. WOOD (1975) "Strands of theory and research in collective behavior," pp. 363-428 in Annual Review of Sociology, Vol. 1. Palo Alto: Annual Reviews.

MELTSNER, A. J. (1971) The Politics of City Revenue. Berkeley: University of California Press.

MEYER, J. R., J. F. KAIN, and M. WOHL (1965) The Urban Transportation Problem. Cambridge, MA: Harvard University Press.

MILLER, W. E. and D. E. STOKES (1963) "Constituency influence in Congress." American Political Science Review 57 (January): 45-56.

MLADENKA, K. R. and K. Q. HILL (1977) "The distribution of benefits in an urban environment: parks and libraries in Houston." Urban Affairs Quarterly 13 (September): 73-92.

MOLOCH, M. (1976) "The city as a growth machine." American Journal of Sociology 82 (September): 50-65.

MONROE, A. D. (1978) "The cultural basis of American politics: testing Elazar's theory." Presented at the Southern Political Science Association meetings, Atlanta, Georgia, November.

MORGAN, W. R. and T. N. CLARK (1973) "The causes of racial disorders: a grievance-level explanation." American Sociological Review 38 (October): 611-624.

MUTH, R. (1971) "Migration: chicken or egg." Southern Economic Journal 37 (January): 295-306.

——— (1969) Citizens and Housing. Chicago: University of Chicago Press.

NATHAN, R. P. et al. (1977) Block Grants for Community Development. Washington, DC: Department of Housing and Urban Development.

National Commission for Manpower Policy (1978) Job Creation Through Public Service Employment. Washington, DC: National Commission for Manpower Policy.

NSF-RANN (1976) Multi-volume series of reports on General Revenue Sharing, including The Economic and Political Impact of General Revenue Sharing. Washington, DC: NSF-RANN.

NEENAN, W. (1970) "The suburban-central city exploitation thesis: one city's tale." National Tax Journal 23: 117-139.

NELSON, W. and P. MERANTO (1977) Electing Black Mayors. Columbus: Ohio State University Press.

NIE, N. H., S. VERBA, and J. R. PETROCIK (1976) The Changing American Voter. Cambridge, MA: Harvard University Press.

NISKANEN, W. (1971) Bureaucracy and Representative Government.. Chicago: AVC.

O'CONNOR, J. (1973) The Fiscal Crisis of the State. New York: St. Martin's Press.

OLSON, M. (1965) The Logic of Collective Action. Cambridge, MA: Harvard University Press.

ORBELL, J. M. and T. UNO (1972) "A theory of neighborhood problem solving: political action vs. residential mobility." American Political Science Review 66 (June): 471-489.

OSTROM, V., C. M. TIEBOUT, and R. WARREN (1961) "The organization of government in metropolitan areas: a theoretical inquiry." American Political Science Review 55 (December): 831-842.

OSTROM, E. and R. B. Parks (1973) "Suburban police departments: too many and too small?" in L. H. Masotti and J. K. Hadden (eds.) The Urbanization of the Suburbs. Beverly Hills, CA: Sage.

OSTROM, E., R. B. PARKS, and G. P. WHITAKER (1978) Patterns of Metropolitan Policing. Cambridge, MA: Ballinger.

OSTROM, E. et al. (1973) Community Organization and the Provision of Police Services. Beverly Hills, CA: Sage.

PAGE, B. I. (1978) Choices and Echoes in Presidential Elections. Chicago: University of Chicago Press.

——— (1977). "Elections and social choice: the state of the evidence." American Journal of Political Science 21 (August): 639-668.

PARSONS, T. (1969) Politics and Social Structure. New York: Free Press.

PATTERSON, S. C. (1968) "The political cultures of the American states." Journal of Politics 30 (February): 187-209.

PHILLIPS, L. and H. L. VOTEY (1977) "An economic basis for the definition and control of crime," pp. 89-109. in S. S. Nagel (ed.) Modeling the Criminal Justice System. Beverly Hills, CA: Sage.

POMPER, G. (1972) "From confusion to clarity." American Political Science Review 66 (June).

RIKER, W. H. (1962) The Theory of Political Coalitions. New Haven, CT: Yale University Press.

——— and P. C. ORDESHOOK (1973) An Introduction to Positive Political Theory. Englewood Cliffs, NJ: Prentice-Hall.

ROBINSON, T. P. and T. R. DYE (1978) "Reformism and black representation on city councils." Social Science Quarterly 59 (June).

RODGERS, B. and C. M. LIPSEY (1974) "Metropolitan reform: citizen evaluations of performance in Nashville-Davidson County, Tennessee." Publius 4 (Fall): 19-34.

ROSSI, P. H. (1979) "Estimating program impacts using time linked small area census data." Presented at the Conference on Comparative Urban Policy Research, Chicago, April.

——— R. BERK, and B. EIDSON (1976) The Roots of Urban Discontent. New York: Wiley-Interscience.

SAMUELSON, P. A. (1969) "Pure theory of public expenditures and taxation," pp. 98-123 in J. Margolis and H. Guitton (eds.) Public Economics. London: Macmillan.

SAVAS, E. S. (1977) The Organization and Efficiency of Solid Waste Collection. Lexington, MA: D. C. Heath.

SCHNEIDER, A. L. (1976) Opinions and Policies in the American States: The Role of Political Characteristics. Beverly Hills, CA: Sage.

SCHNEIDER, M. (1979) "Population change and the fiscal condition of metropolitan communities." Presented at the Conference on Comparative Urban Policy Research, Chicago, April.

SCHUMAKER, P. D. (1979) "Theoretical and Methodological issues regarding the investigation of citizen preferences and policy responsiveness through comparative urban research." Presented at the Conference on Comparative Urban Policy Research, Chicago, April.

———, R. W. GETTER, and T. N. CLARK (1979) Policy Responsiveness and Fiscal Strain in 51 American Communities. Washington, DC: American Political Science Association.

SCHUMAKER, P. D., T. N. CLARK, and R. W. GETTER (1977) "Responsiveness bias in 51 American communities." American Journal of Political Science 21 (May): 247-280.

SCOTT, D. (1976) "Measures of citizen evaluation of local government services," pp. 111-128 in T. N. Clark (ed.) Citizen Preferences and Urban Public Policy. Beverly Hills, CA: Sage.

SEN, A. K. (1970) Collective Choice and Social Welfare. San Francisco: Holden-Day.

SHARKANSKY, I. (1969) "The utility of Elazar's political culture: a research note." Polity 2 (Fall): 66-83.

SHAW, R. P. (1975) Migration Theory and Fact. Philadelphia: Regional Science Research Institute.

SONENBLUM, S., J. J. KIRLIN, and J. C. REIS (1977) How Cities Provide Services. Cambridge, MA: Ballinger.

SPILERMAN, S. (1971) "The causes of racial disturbance: comparison of alternative explanations." American Sociological Review 36 (June): 527-442.

STINCHCOMBE, A. (1975) "Social Structure and politics," pp. 557-622 in F. I. Greenstein and N. W. Polsby (eds.) Handbook of Political Science, Vol. 3. Reading, MA: Addison-Wesley.

SULLIVAN, J. L., J. E. PIERSON, AND G. E. MARCUS (1978) "Ideological contrasting in the mass public: a methodological critique and some new findings." American Journal of Political Science 22 (May): 233-239.

SUTTLES, G. D. (1968) The Social Order of the Slum. Chicago: University of Chicago Press.

SWAAN, A. D. (1973) Coalition Theories and Cabinet Formation. Amsterdam: Elsevier.

SWEET, D. C. [ed.] (1972) Models of Urban Structure. Lexington, MA: D. C. Heath.

TIEBOUT, C. M. (1956) "A pure theory of local expenditures." Journal of Political Economy 64 (October): 416-426.

TOLLEY, G. S., P. E. GRAVES, and J. L. GARDNER (1979) Urban Growth Policy in a Market Economy. New York: Academic Press.

TUCKER, C. J. (1976) "Changing patterns of migration between metropolitan and nonmetropolitan areas in the United States: recent evidence." Demography 13: 435-444.

TURK, H. (1977) Organizations in Modern Life. San Francisco: Jossey-Bass.

U. S. Bureau of the Census (1975a) "Mobility of the population of the United States: March 1970 to March 1975." Current Population Reports, Series P-20, NO. 285. Washington, DC: Government Printing Office.

——— (1975b) "Social and economic characteristics of the metropolitan and nonmetropolitan population: 1974 and 1970." Current Population Reports, Series P-23, No. 55. Washington, DC: Government Printing Office.

VAUGHAN, R. J. and M. E. VOGEL (1979) The Urban Impacts of Federal Policies, Vol. 4: Population and Residential Location. Santa Monica, CA: Rand Corporation.

VERBA, S. and N. H. NIE (1972) Participation in America. New York: Harper & Row.

VERNEZ, G., R. J. VAUGHAN, R. K. YIN (1979) Federal Activities in Urban Economic Development. Santa Monica, CA: Rand Corporation.

WARREN, D. I. (1975) Black Neighborhoods. Ann Arbor: University of Michigan Press.

WARREN, R. L., S. M. ROSE, and A. R. BERGUNDER (1974) The Structure of Urban Reform. Lexington, MA: D. C. Heath.

WEBB, K. and H. P. HATRY (1973) Obtaining Citizen Feedback. Washington, DC: The Urban Institute.

WEICHER, J. C. (1970) "Determinants of central city expenditures: some overlooked factors and problems." National Tax Journal 23 (December): 379-396.

WILLIAMS, O. P. (1961) "Typology for comparative local government." Midwest Journal of Political Science 5 (May): 15-0-164.

——— and C. R. ADRIAN (1963) Four Cities. Philadelphia: University of Pennsylvania Press.

WILSON, J. Q. and E. C. BANFIELD (1971) "Political ethos revisited." American Political Science Review 65 (December): 1048-1062.

——— (1964) "Public-regardiness as a value premise in voting behavior." American Political Science Review 58 (December): 879-887.

WINNIE, R. E. and H. P. HATRY (1972) Measuring the Effectiveness of Local Government Services: Recreation. Washington, DC: The Urban Institute.

WOLFINGER, R. E. and J. O. FIELD (1966) "Political ethos and the structure of city governments." American Political Science Review 60 (June: 306-326.

WRIGHT, G. C. (1976) "Linear models for evaluating conditional relationships." American Journal of Political Science 20 (May): 349-373.

WRIGHT, J. D., P. H. ROSSI, S. R. WIGHT, and E. WEBER-BURDIN (1979) After the Clean-Up: Estimates of Long-Term Disaster Effects. Beverly Hills, CA: Sage.

ZIMMER, B. (1975) "The urban centrifugal drift," pp. 23-91 in A. H. Hawley and V. P. Rock (eds.) Metropolitan America in Contemporary Perspective. New York: Halsted Press.

Part II

Urban Politics and Policy Outcomes

3

Political Leadership and Urban Fiscal Policy

TERRY NICHOLS CLARK
LORNA CROWLEY FERGUSON

THIS CHAPTER includes several subsections, most of which begin with a brief summary of propositions emerging from research to date in the area. Then implications for data collection and analysis are pursued and specific options listed. The greatest amount of space is allocated to those matters which seem either most worthy of data collection or most deserving of clarification in terms of future options. Many research examples are drawn from work using the NORC Permanent Community Sample.[1] An effort has been made to complement the topics addressed by other conference papers.

I. ALTERNATIVE MODELS OF THE POLITICAL PROCESS: CITIZENS, GROUPS, LEADERS

A. THEORIES AND FINDINGS TO DATE

Three somewhat distinct traditions have posited alternative explanations of the political process. The critical independent variables are, respectively, citizens, organized groups, and political leaders. The dependent variable in each case is the policy outputs of the political system, which we here consider primarily in fiscal terms.

Citizen preferences are stressed in the perspective developed by Downs (1957) and his various followers. Organized interest groups are more central in the group theory tradition from Bentley through Key and Dahl (e.g., 1961). The third approach stresses more the specific preferences of political leaders, as in the tradition of elite theories of Mosca, Pareto, and

Michels and continued in many case studies of individual mayors or eminent leaders in American cities (for example, Allen, 1971; Caro, 1974; Rakove, 1975). A variant still stressing the dynamic leader, but in a clearly structured context, is that of the "political entrepreneur" as discussed by Frohlich et al. (1971).

Recently we have formulated an approach that incorporates all three classes of independent variables—citizens, organized groups, and political leaders—but stresses that they are differentially important in varying contexts (Clark and Ferguson, 1979). Two related ideas suggest the conditions under which each type of variable becomes more important. First is uncertainty on the part of political leaders as to citizen preferences. When citizen preferences are unclear, leaders are more likely to rely on organized groups, or their own preferences, in formulating public policy. Second is slack. Citizens may be minimally involved in the political process if they consider public policy to be broadly consistent with their preferences. Within these broad constraints, there is enough slack in the political system for group and leadership preferences to be incorporated. But if policy strays too far from citizen preferences, the slack can be taken up.

The system may thus be considered in equilibrium when policies are consistent with citizen preferences. But as such an equilibrium is readily disturbed, how may we analyze such "disturbances"? It is useful to introduce the concept of "sectors" of citizens, distinguished by their reasonably similar views concerning public policy. The more inconsistent a policy is with the preferences of a given sector, the more that sector is likely to (1) become politically involved, and thus provide support for organized groups representing such policies, or (2) migrate out of the city if more attractive alternatives are available. Or options 1 and 2 may be pursued simultaneously by different members of the same sector.

At the city level, these general propositions suggest that cities *more homogeneous* in terms of citizen preferences are likely to be characterized by lower levels of organized group activity; citizen preferences are likely to be more important determinants of policy outputs. *Political stability* similarly should permit more knowledge to be accumulated by political leaders about citizen preferences and thus lead citizen preferences to play a more important causal role than in times of political instability. Conversely, the less homogeneous the city and the greater the political instability, the more important should organized groups and political leaders' preferences be in influencing policy outputs.

These propositions are generally consistent with our analysis of urban fiscal policy from 1960 to 1977 in 62 of the Permanent Community Sample cities. Sector size measures were created for three sectors: the poor (percentage of residents below the federal level of "low" income in 1970); blacks; and middle-class citizens (usually the percentage of families with incomes from $10-15,000 in 1970). These specific sectors were chosen in conjunction with extensive reviews of the literature on citizen policy preferences and analysis of several surveys for individual citizens (discussed below). Organized groups and leaders' preference measures were constructed using survey data, as will be discussed further below. The fiscal policy outputs were levels of expenditures and debt per capita in 1977 and percentage changes in these for several time periods from 1960 to 1977.

A regression model was specified, including among the independent variables sector size, organized group activities, and political leaders' preferences; and the model was estimated to explain fiscal policy in several time periods. In brief, we found that the sector size measures, particularly the middle-class sector, was most critical in the early 1960s and the mid- to late 1970s. By contrast, political leaders' preferences were more important in the middle years: late 1960s and early 1970s. And in a recursive analysis, political leaders' preferences in these middle years were in turn largely explained by organized group activity, more than by sector size. Scatter plots were constructed to array the cities in a two-dimensional space in terms of the most critical independent variables for each time period (sector size or leaders' preferences tabulated against change in fiscal policy). This permitted us to isolate individual cities illustrating the national trend to varying degrees. In all three periods, the smaller, more homogeneous cities seem to maintain slow but consistent increases in expenditure. In the middle period, those with the most active organized groups and mayors who prefer the highest spending increase expenditures the most. Different processes are thus important in different cities and time periods.

This admittedly synoptic summary is included to suggest the interesting possibilities offered by comparative urban research for extending some of the core theories in the social sciences. If we had tested each of the three hypotheses about citizen preferences, organized groups, or political leaders in an individual city at one point in time, we might have found mixed support. It is not likely that we would have been led to elaborate an interpretation explicitly introducing contextual characteristics to explain the differential importance of the various explanatory variables. Only the availability of comparable data over time for a large number of cities permitted the kind of analysis necessary to test such a theory. But we do not wish to imply pure deduction as the process leading to these ideas; it has been immersion in comparative urban data over many years that led us to the contextual propositions.

B. DATA IMPLICATIONS

We address implications of such propositions for empirical work in terms of the three classes of variables: citizens' preferences, organized groups, and political leaders.

(1). Citizen Preferences. In the analysis just summarized, we followed an approach frequently used in the last decade: analysis of census data concerning socioeconomic characteristics of the population to infer public policy preferences of citizens. Economists such as Davis have followed such a procedure with few explicit reservations as to the adequacy of such data as indicators of citizen preferences. Indeed, many economists work from a revealed preference conception which easily leads to the view that whatever the political system produces constitutes the revelation of citizen preferences. This assumes the workings of a market-like situation where political leaders become invisible in translating citizen preferences into public policy. Our analysis suggested that this may in fact be reasonable for certain cities and time periods, but it remains an issue quite amenable to more direct empirical analysis. Several approaches are possible, at varying costs.

(a) Use census data concerning the size of various "sectors." This can be pursued not only in isolation, but combined with other approaches. It remains the most straightforward and inexpensive method, as data are available in the *County and City Data Book,* and related census reports, for classifying the percentage of citizens by standard socioeconomic characteristics in various cities. Ethnicity and religion, however, are particularly poorly represented by census data; policy outputs where these are likely to be important are thus less amenable to study using this approach.

(b) Collect and analyze data concerning citizen preferences using standard national samples of the U.S. population. This approach is particularly inexpensive if data are already available and secondary analysis is possible. However, remarkably few survey items have addressed *urban* public policy questions; the great majority concern national issues such as inflation, war, income taxes, and the like. Such items have been collected by Gallup, Harris, and Roper off and on since the 1930s, and more systematically by SRC and NORC, especially in NORC-GSS surveys from 1972 onward. Most GSS public policy items, however, and all expenditure policy items, concern *national* policies. There are a few exceptions in other NORC surveys, such as one of housing by Butler et al., and some NORC-CNS surveys. The Michigan election surveys have recently added a range of local participation and policy-related items, but these have concerned almost exclusively support for and involvement with U.S. congressional candidates rather than urban candidates or

issues. Also, the GSS is adding contextual data on local area characteristics from census-type sources. It is good to see movement in this direction, but it still leaves much to be desired from the standpoint of urban public policy.

Consequently, researchers concerned with analyzing citizen preferences on urban public policy have turned to other studies, not because urban policy items cannot be asked in national surveys, but because they seldom have been. The best studies of citizen preferences concerning urban public policy seem to be the Urban Observatory study of about 500 citizens in each of 10 cities, conducted in 1970 (see Fowler, 1974; Hoffman, 1976) and studies of citizens in individual metropolitan areas (for example, Scott, 1976; McIver and Ostrom, 1976).

The limitations of such individual city surveys are clear, but the standard national survey of some 1500 respondents also presents difficulties. One obvious problem is that the number of cases in each primary sampling unit (PSU) drops to 15 or so. Almost equally important, PSUs are defined by most survey organizations as metropolitan areas, and respondents are not coded by specific city residence. The number of cases increases dramatically using a repeated survey design like the CNS or GSS (to about 8000). Hence one possibility is to add to such surveys certain urban-type questions. Still, just what changes in sampling procedures and documentation would be possible is not fully clear. Even if the number of cases remains low in a single city, analysis conducted across a large number of cities using both citizen and city-level data could still generate suggestive results. This point was made by Rossi and Crain (1969), but efforts we have made to pursue this approach have run into the problem of documentation of respondents only according to metropolitan area (Clark and Nanda, 1972). For items like satisfaction with police or public schools, or urban leadership, this is disastrous.

(c) Simulate citizen preferences in numerous cities using data from surveys in other cities. This approach, developed by de Sola Pool, Ableson, and Popkin (1965), has been applied to citizens in states by Shaffer and Weber (1974) and in cities by Schumaker and Getter in several papers (see the Schumaker paper in this volume). The basic idea is to cross-classify citizens by social background characteristics, thereby generating numerous subcategories (for example, white women over 65 with incomes below $6,000 residing in single-family dwellings in cities from 50,000 to 100,000). The response scores on various questionnaire items are generated for each subcategory, in the case of Schumaker and Getter, using the Urban Observatory survey data. The census-type data from cities are used to specify the number of citizens who fall into each subcategory of the kind used in the questionnaire. A city-level simulated score for a questionnaire item is then generated by summing the

responses to that item by each subcategory of respondents weighted by the size of that subcategory in the city.

This procedure builds on the assumption that relationships between social background characteristics and citizen policy preferences do not vary across cities or policy areas. Hoffman and Clark (1979) have analyzed the same Urban Observatory data to assess the size of respondent, neighborhood, and city-level effects by different policy areas. We found that the variance explained by each class of variables differed considerably by policy area. As much as 86 percent of the variance explained is by individual-level variables for attitudes toward spending on welfare services, but only 44 percent for environmental services such as air pollution. (Neighborhood and city-level characteristics account for 56 percent of the explained variance for environmental services.) Nevertheless, Schumaker and his associates have compared the results of the simulation approach with direct evidence from the Urban Observatory surveys and found strikingly similar results using the two procedures.

(d) Survey citizens in one or a small number of cities. Studying a single city using 1000 or so respondents, or as many as four cities differing dramatically on certain variables likely to affect citizen preference patterns, is a simple approach. Especially for exploratory purposes—for new public issues or refining new kinds of questionnaire items—this seems especially useful to consider.

Another rationale for a limited number of cities is where several levels of detailed data collection and analysis are called for, such as in the Rossi et al. (1974) study of service delivery by police, social workers, public school teachers, and small merchants. Data were collected from citizens using these services, from the service providers (policemen, and so on), and from city leaders in 15 cities. Effects of variables at all three levels were thus distinguished.

(e) Reduce costs using mail and telephone interviews. Data collection costs drop dramatically if mailed questionnaires or telephone interviews are used in lieu of personal interviews. Response rates to mailed surveys tend to remain around 50 percent, even with two or three mailings, thus posing sampling problems. But telephone interviews seem to generate data quite comparable to those from personal interviews using most questionnaire items (see Sudman and Bradburn, 1974). Telephone interviews cost about $30 to $50 each for a 20-30 minute interview, while personal interviews are closer to $150. A mailed survey is about $20 per case. These are recent NORC estimates including some pretesting, questionnaire printing, interviews, coding, punching, and preparation of clean tape and codebook.[2]

(f) Questionnaire formats for public policy preferences. This has been a subject of some controversy. Traditional poll items with yes/no response categories were criticized as not including intensity of preference. Likert-type continuous items, such as five-point scales from agree strongly to disagree strongly, have thus been used in an effort to gauge intensity of preference. But this format still does not impose a "budget constraint" analogous to the limits on city spending on each item implied by the size of the total budget. The budget pie was thus developed to incorporate both intensity of preference and a budget constraint. Formats varied from a circle which the respondent was asked to divide into pie slices, to coins he could place in different cups, to points he could assign to various expenditure areas. A good deal of interesting theoretical work has been done in the public choice tradition which suggests the desirability of budget pies and related approaches to assessing citizen preferences for public goods. (See the overviews in Clark, 1974a, and Tideman, 1976.) However, many of these theoretical demands seem too arduous for average citizens. A few methodologically interesting studies have been conducted which compare results from similar items asked using card sorts, ordinal scales, self-anchoring scales, and budget pies (Scott, 1976; McIver and Ostrom, 1976). Although for some specialized purposes the more refined formats have advantages, difficulties of administration have led many researchers to prefer simpler formats such as "Do you feel that more, less, or about the same amount should be spent on (service)"? This format is used in many CNS and GSS items and the Urban Observatory study.

(g) Political culture and participation of citizens. Surveys of citizen policy preferences seem quite appropriately joined in data collection efforts with political culture items. Indeed, political culture can refer both to policy preferences and more general views about norms and institutions legitimately influencing the political process. Similarly, certain self-reported behavioral measures, such as voting turnout, registration, campaign activity, and support for candidates or referenda, could be simultaneously collected from citizens.

(h) Citizen preferences and activities from nonsurvey sources. Apparently only one national study has ever been conducted collecting data concerning voter turnout of citizens in American local elections (Alford and Lee, 1968), and no one has systematically collected or monitored voting data by party in cities. Such data are routinely collected and published for congressional and presidential elections by Richard Scammon and others. But these efforts have always left out municipal elections. This is a reasonably low-cost area of data collection that could yield interesting benefits. Probably the local city clerk would be the best

source, although local newspapers are a possibility, especially for the past. Election results take on special interest when available over many years. Probably mayoral and council elections are the most critical. Turnout and percentage of eligible voters registered would also be interesting, but these may not be maintained in comparable manner across cities. If the number of voters is available, however, this can at least be compared to the size of the adult population of the city.

Such data may appear mildly boring for most cities most of the time. But in situations like those that existed in Cleveland and Gary in the late 1960s, when blacks were being mobilized, the white turnout increased as well, with 80 and 90 percent of those eligible voting, according to local reports. This would indicate a taking up of much of the slack in the political system and very possibly a shift in the relative importance of the different actors involved. How much the taxpayers' revolt of the late 1970s can be discerned using such data would be fascinating to analyze. Many PCS cities were cutting back in spending and the number of municipal employees from 1974 to 1976, four years before Proposition 13 (in 1978). But leaders in many such cities reported that they favored increased spending in the 1976 survey. How were cues from voters transmitted?

It would be interesting to link the mobilization of voters to their varying degrees of impact on political leaders and public policy over time. The major activities of blacks in the late 1960s and early '70s have been documented with a few case studies, but only scant comparative analysis. Similar patterns involving Spanish-speaking persons have been under way in many cities and succeeded in quite different manner in different places (for example, San Antonio versus Houston). How these earlier patterns link, if at all, with more recent middle class reactions deserves study.

(2) Organized Groups. These include downtown business associations, municipal employees, neighborhood associations, civic groups, and others often involved in local decisions. Variations of positional, reputational, and decisional approaches have been used to study the import of such groups in cities. Case studies of individual cities abound, but systematic comparative research is more limited. The 1967 Permanent Community Sample study included all three types of items and thus provided baseline measures for comparison with subsequent studies.

Organized groups may be analyzed individually, following such basic propositions as the more the resources at an organization's disposal, the greater its reputational power, the greater its political activity; and the more it favors a given policy, the more likely is that policy to be approved. The same basic idea may be restated for all organizations in the city: the more organizations that favor a policy and the greater their resources, reputational power, and political activity, the more likely the policy is to be approved.

These propositions treat organizations in additive terms; they could be restated to incorporate the degree of consistency with which all local organizations favor a goal, and the degree to which they are interrelated and have worked together and with the city council in other matters. Such propositions emerge from the traditions of spatial modeling, coalition theory, and networks, addressed in the paper by Burt.

Four types of data concerning organized groups seem important to collect. First is *resources*. Size and social background status of membership, budget, size of professional staff, and salary of professional staff are standard internal organizational characteristics easy to list but very seldom collected on a comparative urban basis. External or interorganizational characteristics include number and intensity of overlapping memberships of leading organizational members with other important institutions, as emphasized in recent work on networks.

Second is *reputational power,* such as tapped by reputational-type items, preferably "issue-specific" versions. For example, "If someone wanted to be elected mayor in (city), is there any individual or group whose support would be essential?" This was our "open-ended issue-specific reputational" item used in 1967; it was repeated for four other issue areas aside from the mayoral election: urban renewal, school board elections, issuing municipal bonds, and air pollution control. The same basic question was also used in a "closed-ended issue-specific reputational" format that included a list of 15 organizations which the respondent was asked to score as "essential," "important but not essential," or "not important." Approximately similar items have been used in several subsequent surveys with the PCS, most recently in 1977.

The third type of data is *political activities* or *influence* as gauged by actual activities of the organization aimed at influencing public decisions. These matters have been studied with a variety of procedures that may be labeled *decisional*, such as a series of questions that inquire "Who initiated, who supported, who opposed, who mediated, and who prevailed?" in a specific decision such as the last largest urban renewal project in the city. Such items were used to study air pollution, urban renewal, selecting the mayor, and antipoverty programs in the PCS cities in 1967. (See Clark, 1975b, for details on formats and related matters.) Variations have been used in PCS resurveys, such as the 1976 survey administered by Schumaker and Getter. They asked mayors and city councilmembers simply to list the frequency with which they were contacted by the five most important organizations in the city, and then the frequency with which they contacted the organizations.

Data from nonsurvey sources that can be used in broadly similiar manner include stikes by municipal employees (available from the Census Bureau), contributions to political campaigns by various organizations (now often public), and numbers of voters registered in registration drives.

The fourth data type concerns *goals* of the organization, or at least policy preferences of their leading members. There has been some theoretical debate in recent years in this area, the traditional concept of organizational goal being criticized by Coleman and others as too anthropomorphic. Related conceptions such as interests or policy preferences of leading members of the organization may be more amenable to direct measurement.

The importance of direct data concerning goals has not been adequately recognized until recently in urban research. The issues emerge forcefully in a study by Zimmermann (1979) which computed measures of the degree to which various local organizations favored greater assistance to the poor. What was remarkable was the degree of difference in organizational goals across cities. Data were available for the same nine types of organizations—among them, Democratic and Republican parties, businesses, newspapers, labor unions, and civic groups. In Amarillo, *all* of these organizations were opposed to helping the poor, while in Malden, Massachusetts, *all* favored helping the poor. In other cities there was considerable variation; for example, in Palo Alto, the Republican party favored helping the poor. Such dramatic differences underscore the dangers of imputing goals to organizations and assuming they are comparable across cities, as in Morlock's (1974) characterization of business and civic groups as conservative. A quick review of the literature has not disclosed a single additional comparative study that has documented or analyzed differences in general goals of local organizations, although we do have measures of participation in favor of or opposed to various specific issues in our PCS decisional items for 1967.

Data of all four types outlined above may be collected most simply by using a standardized panel of informants, identical in all cities, to maximize informant reliability. Examples are chief administrative official of the city, finance director, chairman of the city council finance committee, city bureau chief of the largest newspaper, head of the largest municipal employees union or association, and others depending on the specific focus of the study.

(3) Political Leadership. Some basic hypotheses include the following: The more leaders favor a given policy, the more likely it is to be implemented. The more particularistic the leader, the more impact

neighborhood associations and individuals should have on public policy. The more universalistic (or "unitary" or "reform-oriented") the leader, the more impact citywide organizations will have on public policy. The more universalistic the leader, the higher the level of public goods (as opposed to separable goods) provided by the city.

Another new hypothesis concerns the New Fiscal Populist (Clark and Ferguson, 1979). Leaders like Peter Flaherty or Jimmy Carter are (1) new, (2) fiscally conservative, (3) socially liberal, (4) populist in appealing directly to citizen-voters without relying on parties or other organizations often important in the past. They are also usually (5) Democrats, but differ from the New Deal Democrats by their fiscal conservatism and greater populism. Such leaders are emerging in many cities as well as in the U.S. Congress (see Schneider and Schell, 1978). This new ideological pattern conforms to none of the traditional Democratic or Republican models, or machine or reform types, but combines elements of all. It represents the most profound shift in policy orientation of urban leaders since the New Deal and is one with dramatic consequences in terms of fiscal retrenchment. A good deal more work is called for to document the emergence of New Fiscal Populists as mayors and city councilmembers across the country, and in turn to trace their impacts on public policy.

Certain data have been collected using the PCS concerning such leadership characteristics. The 1967 survey included leaders basic social background data and certain activities, but few value items. Value items were stressed in the International Studies of Values in Politics, conducted in 30 cities. The 1976 PCS survey by Schumaker and Getter included items concerning social background, policy preferences, and activities. Given continual changes in these matters, monitoring leaders on some core items seems particularly important. If certain items could be replicated from these past surveys, this would permit systematic comparison over time. Interview costs are approximately the same as those of citizens, indicated above.

Some basic data that might be collected are these:

(a) Social background characteristics. Age, education, father's occupation, religion, ethnicity, race, party, occupational positions held past and currently.

(b) Values and policy preferences. Political culture items concerning norms and behavior appropriate to municipal public affairs (legitimacy of public versus separable goods, localism, particularism, and so on; see the Knoke paper). Policy preferences concerning more, less, or the same level of spending in 11 basic functional areas: among them, education, police, and fire. (This item was used in the 1976 PCS survey.)

(c) Activites. Here basic decisional data might include certain critical recent policies, such as whether there were significant changes in the

budget in recent years, and how they were incorporated. Decisional items could be used similar to those for organized groups as discussed above. Frequency of contact with various types of organizations might also be included.

II. LEGAL STRUCTURES AND PRIVATE SECTOR RESOURCES

A. THEORETICAL PERSPECTIVE

The propositions about citizens, groups, and leaders are phrased in ceteris paribus manner to simplify exposition. Two particularly important sets of variables that deserve consideration besides those outlined above are legal structures and private sector resources. Both are best conceived not as exerting direct effects on fiscal policy, but as interacting with the above variables to facilitate or retard the effect. For example, the more political leaders favor a given policy and the more legal authority they have, the more likely are policy outputs to be consistent with their preferences. Further, the more political leaders favor a given policy, and the more private sector resources are at their disposal, the more policy outputs are likely to be consistent with their preferences.

Another, relatively understressed, aspect of government structure is the part played by nonelected municipal officials or "bureaucrats" in the emerging theoretical literature (Niskanen, 1971). Bureaucrats can shape or reshape policies in line with their own preferences. Most research on this to date has focused exclusively on the national level of government (but see the Parks and Ostrom paper).

B. DATA IMPLICATIONS

Most data in these areas are reasonably easy to access and generally amenable to reinterpretation from published sources, with a few exceptions. Many such variables are currently being analyzed by researchers. So perhaps quite limited and reasonably inexpensive data acquisition on selective issues is most appropriate here.

(1) Reform Government and Legal Arrangements among Local Governmental Officials. These basic data are collected periodically by the International City Management Association and either published in *The Municipal Yearbook* or made available on computer tapes (for example, GOV74) for individual cities. There are occasional problems of missing data and inadequate coverage of certain issues, such as civil service, but the data seem appropriate to sustain most current theorizing in the area by social scientists. Separate characteristics as well as summary measures

have been constructed and analyzed, such as indexes of reform government, mayoral authority, council authority, and authority of all elected officials (see the symposium in *Social Science Quarterly,* June 1978, Clark and Ferguson, 1979).

(2) Local Autonomy. Here the data are less precise and many conceptual problems remain (Stonecash, 1978; Turk, 1977). Several ACIR publications list characteristics, generally for all cities in a state, but without detail on individual city variations or interpretations of the legal structure by state and local administrative agencies. Still, the ACIR data are collected by survey, and to improve on this would be quite costly. This area may be reasonable to consider if there is special interest in a particular issue, but it probably does not warrant top priority at present. Measures have been assembled for types of *debt limits, revenue structure and expenditure limits, overall fiscal monitoring, metropolitan government,* and *federal- and state-mandated activities,* especially concerning public employees. (These are reviewed in Clark and Ferguson, 1979, and MacManus, 1978.)

(3) Functional Performance. Perhaps the most critical single variable in this area for fiscal policy outputs is the range of functions performed by the city government, as made clear in several studies by Liebert (especially 1976). Twelve different methodologies for addressing the problem are outlined and compared in Clark et al. (1979); but the matter demands continuing attention, first, because cities change their functional responsibilities rather often, commonly by adding subfunctions or transferring activities to special districts. This is enough to impair data for time series comparisons. Eight of our 62 PCS cities changed functional responsibilities from 1960 to 1974, many quite substantially. Second, the range of functions performed by a city apparently interacts with other decision-making characteristics in ways not yet clarified. For example, municipal employees seem more unionized and salaries are higher in cities performing more functions, even after controlling several related variables.

Research in this area might take two forms. First is collection of data concerning the specific functions performed in a sample of cities. Certain local documents, such as financial reports, could be consulted in conjunction with interviews of local officials about changes in functions over the last decade or so. Then more precise scores concerning the specific functions performed could be computed over time. These could then be used in comparative analysis of changes in expenditures. Second, more case study work in a limited number of cities concerning the causes and effects of changes in functional responsibility would seem useful, as very little such work has been completed.

(4) Private Sector Resources. These include median family income, taxable property base, and the like, available from census-type sources.

(5) Bureaucrats. Procedures used to study policy preferences of elected officials could similarly be applied to bureaucratic officials. See the work on these matters in Friedman (1975), Levy et al. (1974), and Borcherding (1977).

III. POLICY OUTPUTS, ESPECIALLY FISCAL

A. THEORETICAL WORK

The extension of the research question "Who governs?" to "Where, when, and with what effects?" has been fairly general. But on how to answer these last questions there is less consensus, although almost all solutions involve some consideration of policy outputs. Classification of policies such as redistributive, regulative, allocative, public good, and separable good is a beginning. Linking these various types of policy output to distinctive patterns of decision-making is a next important theoretical step. Propositions like centralization encourage public goods, but decentralization generates separable goods can thus be formulated (see Clark 1975a). Under somewhat different headings, related matters are discussed by sociologists, political scientists, public finance and public choice economists, and public administration theorists (examples are Grimes et al. 1975; Smith, 1979; and Musgrave, 1959).

On the specific topics of fiscal policy and fiscal strain, more theoretical work is very much in order. Drawing systematic linkages between municipal financial performance and bond market performance is one avenue to pursue here. More careful analysis of relations between various types of funding and revenue structures (local and intergovernmental), and expenditure and debt policies is also clearly in order. The theoretical discussions of such issues as incrementalism, substitution, and stimulation have developed without explicit linkage to the sorts of decision-making processes outlined above. These several literatures could be strengthened by development of overlapping themes. Better data are essential to address many of the theoretical questions.

B. DATA IMPLICATIONS

Over the last decade fairly general consensus has emerged concerning the importance of linking urban leadership and decision-making to specific policy outputs. Nevertheless, specific types of output have been debated, and several alternatives pursued. *Fiscal policy outputs* have been studied because of the importance of money and the ready quantification of policies in different functional areas. *Performance outputs,* such as tons of garbage collected, have been analyzed in some outstanding case studies (for example, Lineberry, 1977), but there are few comparative studies, given the difficulties of data collection. *Policy impacts*, such as greater satisfaction among citizens due to improved garbage collection, are even more difficult to measure. All three classes of variables are important and deserve further work, but given the complications of collecting and interpreting meaningful data concerning many performance and impact measures, for the moment it may be most appropriate to continue work on these using primarily research in limited numbers of cities.

One criticism of fiscal policy outputs in the late 1960s was that they were not "relevant" enough. This has changed in a time of fiscal austerity and fiscal strain for many cities. Since the1975 New York City fiscal crisis, social scientists have been increasingly asked to help interpret these developments. The paucity of earlier work has become clear, and the importance of developing a more refined set of indicators of fiscal policy outputs and fiscal strain has grown among users as well as producers of research in the area. The poorly documented and often incomparable nature of financial reporting by individual cities impairs too detailed work, but this is not about to change unless major national forces alter the situation. To make comparable city-level data, the standardizing activities of the Governments Division of the U.S. Census Bureau are an essential preliminary for most comparative urban research. Nevertheless, many of their procedures remain underdocumented. The resulting data have often been criticized but hardly ever subjected to methodological analysis to determine the magnitude of variation in reporting related to the different procedures used by individual cities and the Census Bureau (for example, due to accrual versus cash basis accounting, netting out of interfund transfers, allocation of intergovernmental revenues to different functional areas, and fringe and retirement benefit reporting). It would be useful to attempt estimates of the degree to which such matters affect various fiscal data (for example, welfare or public hospitals data may be less comparable across cities than general expenditures). Procedures used by social scientists to decompose measurement error could be productively applied in this area, probably by recoding data stored in archives of the Governments Division of the Census Bureau (see Clark, 1977b).

Several categories reported by the census could be productively reclassified for many analytical purposes. For example, intergovernmental revenues could be broken down by degree of restrictiveness (block grants versus categoricals of varying specificity). Short-term debt should be disaggregated; in particular, housing and urban renewal bonds guaranteed by the federal government should be listed separately. Changes in functional responsibility, as discussed in the last section, should be documented. These and similar matters could be included in publications by the Census Bureau or ancillary documents prepared specifically for analytical purposes.

The *Annual Survey of Governments* is distributed as a computer file by the Census Bureau but formated so that cities are ordered within states. ICPSR reformated the 1973 and 1974 files, but has not prepared subsequent files. It would be useful to prepare and distribute these data.

IV. HISTORICAL AND COMPARATIVE PERSPECTIVES

Much urban research has focused on American cities in recent years. This is understandable given user concerns for policy implications, student interest, and the quality and accessibility of recent data. Still, to help interpret many of the broader structural issues concerning American cities, comparisons over a longer historical period, and with certain non-U.S. cities, can yield important insights. In some instances, this may reinforce the degree to which current problems seem similar to those of other times and places, as in the Newton et al. analyses of European fiscal difficulties (presented in draft papers at the ISA Uppsala meetings), or in Bryce's depiction of New York's fiscal crises and political leadership problems in the late nineteenth century. However, several broader theoretical efforts have remained quite abstract, partly due to the absence of available data (Przeworski and Teune, 1970; Clark, 1974b; Friedland et al., 1977).

Suggestive historical work could be completed using past data for U.S. cities, following the important, but limited, examples of the Hollingsworths (1979), and Liebert (1976). They have added useful insights into current analyses of expenditures using a variety of social, economic, legal, and political variables from the past. These directions could be encouraged at minimal cost. Some of the new urban historians, as well as historically oriented social scientists, are interested in such efforts. A good deal of data are available from the U.S. Bureau of the Census which could be prepared in machine-readable form at reasonable cost. Perhaps 100 variables concerning basic fiscal, social, economic, and legal characteristics of the 50 or so cities reported by the Census Bureau from about 1850 onward would be worth treating in this way.[3] It would be useful to consult with some of the social science oriented historians through the professional associations in this area. Such issues as the roles of immigrant groups, economic development and diversification, adoption of reform government structure and modifying functional responsibilities of city government, changing income and other social characteristics of citizens over time in conjunction with migration can be studied productively with data for the last century.

International comparisons are more complex, involving greater difficulties in assembling comparable items and interpretation. Several initiatives have been launched in this direction (International Studies of Values in Politics, the Volkswagen-ECPR project of Newton et al., Aiken's summary of data archives, Council for European Studies, and numerous conferences). The journal *Comparative Urban Research* has reported many of these. Researchers working on overlapping problems in different countries try to keep in touch with one another, circulating papers and occasional data. But differences in theoretical perspectives, complications in interpreting legal and political arrangements across national boundaries, and high costs of international travel have discouraged many of these efforts. In the longer term such work is doubtless important and likely to yield useful insights, but the administrative problems involved suggest that for the moment such initiatives may best be left to individuals or small teams of researchers.

V. ORGANIZATIONAL QUESTIONS

A. PREPARATION AND DISTRIBUTION OF DATA

Some of the major studies by social scientists, and major census-type sources, have been assembled, cleaned, and distributed through ICPSR.[4] Many have been simply distributed as received, others cleaned and substantially reformated. Costs thus vary according to the degree of programming and related activities. Some activities, however, must be completed by individual researchers, to prepare and document data sets enough even to turn over to ICPSR.

Tom Anton has a project involving urban fiscal data that deserves consideration of the division of responsibility between him and ICPSR for such activities in future years. In particular, if it would be possible to obtain basic data from the *Annual Survey of Governments* and similar sources more rapidly, then researchers would not have to prepare these individually.

B. DATA COLLECTION

NORC has been involved in several waves of city-level data collection using the Permanent Community Sample, for which several thousand variables are now available in computer files. Individual researchers have frequently used the PCS sampling frame for adding further data for their own purposes to those already available. Whatever future efforts concerning data collection are pursued, it would be useful to try to use sampling frames which overlap as much as possible with existing data sets to permit analyses of change on critical variables.

The PCS was established using the Primary Sampling Units of NORC, thus assuring representativeness of the places of residence of the American population (Rossi and Crain, 1968). Initially, 200 PSUs were used, and subsamples drawn depending on the size of the budget for various projects. Hence surveys of 100, 62, 51, or 30 cities have been used since 1967. Given the multicollinearity problems that plague most social science data, including comparative urban data, and the importance of considering a number of causes of any given phenomenon, a sample of about 50 seems as low as is reasonable to be able to specify models with up to about 10 independent variables.

The number of cases sampled will depend heavily on considerations of cost and data availability. We have developed a three-tier approach—census-type data, surveys, and case studies—that maximizes the total sample, whenever possible.

We have often studied simultaneously two to four individual cities in greater depth, collecting data using more personal interviews, participant observation, and numerous nonsurvey sources in conjunction with PCS analyses. Selection of specific cities has depended on the principal variables being analyzed, with selection often made after completing some comparative statistical analysis to permit selection of extreme cases on, as well as deviant cases off, the regression line.

In assembling data for the PCS, we have acquired many tapes from the Census Bureau and other sources, such as *County and City Data Books* and *Annual Survey of Governments* which we keep on file in such manner that we can access these for occasional analyses where it is unnecessary to add any original survey data. We then have the statistical benefits of using an n of 481 if all cities over 50,000 are included. However, when original data are important to collect from cities through surveys, or simply to code and punch several variables from a variety of public sources, the time and costs are much higher with an n of 481 than one of 62 or 51.

The implications for futher work are that when compiling census-type source materials for which data are on tape for all U.S. cities over 25,000 or 50,000, it may be reasonable to retain the complete data file. But when original data have to be collected or added by hand from several sources, an n of from 50 to 100 is often reasonable. For example, a low-budget telephone interview survey of two informants in 50 cities would cost about \$3,000 (50 x 2 x \$30/interview); ten leaders would be similarly interviewed in 50 cities for \$15,000 (50 x 10 x \$30). (Costs of mailed surveys and personal interviews are discussed in the section above on citizen preferences.)

C. DATA ANALYSIS

Most data analysis has traditionally been completed by individuals or small teams of researchers. As long as data can be distributed for little more than the cost of duplicating magnetic tapes and documentation, this seems likely to continue. Still, occasional collaborative efforts have been assisted by summer workshops, pre- and postdoctoral fellowships, and small grants to permit researchers to visit at certain locations.

D. CONFERENCES AND PROFESSIONAL MEETINGS

An extension of data analysis activities is presentation and discussion of research reports at conferences and professional meetings. A number of overlapping national and international professional associations help maintain communication in the comparative urban area. The smaller, specialized working conference can also be a useful ancillary to ongoing research. Whether funds should be pursued to support such activities, or whether they might better be allocated to data collection and dissemination, are worthy subjects for discussion.

NOTES

1. Clark (1977a) is an overview of work in progress by many researchers using the PCS. Our draft volume Clark and Ferguson (1979) elaborates many points made briefly in this paper. Clark (1975b) is a presentation of the sampling frame and basic leadership items. Schumaker et al. and Clark (1979) is a manual accompanied by a set of selected variables for teaching purposes.

2. These figures were based on studies using a standardized list of respondents (for example, mayor or finance director). Special sampling procedures would increase costs. Costs are approximately the same for citizens and elite respondents.

3. Apparently at least three separate efforts have been made to initiate this work (at Chicago, UCLA, and SUNY Albany), but to our knowledge none has been completed.

4. A list of urban data sets available through ICPSR was prepared in Clark (1975c).

REFERENCES

ALFORD, R. R. and E. C. LEE (1968) "Voting turnout in American cities." American Political Science Review 62: 796-813.

ALLEN, I., Jr. (1971) Mayor: Notes on the Sixties. New York: Simon & Schuster.

BORCHERDING, T. E. [ed.] (1977) Budgets and Bureaucrats: The Sources of Government Growth. Durham, NC: Duke University Press.

CARO, R. A. (1974) The Power Broker: Robert Moses and the Fall of New York. New York: Alfred A. Knopf.

CLARK, T. N. (1977a) "Research in progress using the Permanent Community Sample." Comparative Urban Research 5: 60-71.

——— (1977b) "Fiscal management of American cities: funds flow indicators." Journal of Accounting Research 15: 54-106.

——— (1975a) "Community power," in A. Inkeles (ed.) Annual Review of Sociology, Vol. 1. Palo Alto: Annual Reviews, Inc.

——— (1975b) Cities Differ—But How and Why? Inputs to National Urban Policy from Research on Decision-Making in 51 American Municipalities. HUD Report No. RES 1310. Washington, DC: Government Printing Office.

——— (1975c) Archival Data for Community Teaching and Research: An Introductory Comment. Ann Arbor: ICPSR.

——— (1974a) "Can you cut a budget pie?" Policy and Politics 3: 3-32.

——— [ed.] (1974b) Comparative Community Politics. New York: Halsted Press.

——— and L. C. Ferguson (1979) Political Leadership and Urban Fiscal Policy. (unpublished)

CLARK, T. N. and K. NANDA (1972) "Citizen demands and public policy: a preliminary report." Presented at the annual meeting of the American Political Science Association, September.

CLARK, T. N., L. C. FERGUSON, and R. Y. SHAPIRO (1979) "Functional performance analysis: a new approach to the study of municipal expenditures." (unpublished) To appear in Political Methodology.

DAHL, R. A. (1961) Who Governs? New Haven, CT: Yale University Press.

de SOLA POOL, I., R. P. ABLESON, and S. POPKIN (1965) Candidates, Issues and Strategies. Cambridge: MIT Press.

FOWLER, F. J., Jr. (1974) Citizen Attitudes Toward Local Government, Services and Taxes. Cambridge, MA: Ballinger.

FRIEDLAND, R., F. F. PIVEN, and R. R. ALFORD (1977) "Political conflict, urban structure, and the fiscal crisis." Presented to the American Sociological Association, Chicago.

FRIEDMAN, L. B. (1975) Budgeting Municipal Expenditures. New York: Praeger.

FROHLICH, N., J. A. OPPENHEIMER, and O. R. YOUNG (1971) Political Leadership and Collective Goods. Princeton, NJ: Princeton University Press.

GRIMES, M. D., C. M. BONJEAN, J. L. LYON, and R. L. LINEBERRY (1976) "Community structure and leadership arrangements: a multidimensional analysis." American Sociological Review 41: 706-725.

HOFFMAN, W. L. (1976) "The democratic response of urban governments: an empirical test with simple spatial models." Policy and Politics 4: 51-74

HOLLINGSWORTH, J. R. and E. J. HOLLINGSWORTH (1979) Dimensions in Urban History. Madison: University of Wisconsin Press.

LEVY, F., A. MELTSNER, and A. WILDAVSKY (1974) Urban Outcomes: Schools, Streets, Libraries. Berkeley: University of California Press.

LIEBERT, R. J. (1976) Disintegration and Political Action: The Changing Functions of City Governments in America. New York: Academic Press.

LINEBERRY, R. L. (1977) Equality and Urban Policy: The Distribution Of Municipal Public Services. Beverly Hills, CA: Sage.

LOWI, T. (1968) Article on policy arenas in World Politics.

LYONS, W. et al. (1978) Symposium on reform government. Social Science Quarterly 59 (June).

MacMANUS, S. A. (1978) Revenue Patterns in U.S. Cities and Suburbs. New York: Praeger.

McIVER, J. P. and E. OSTROM (1976) "Using budget pies to reveal preferences: validation of a survey instrument," pp. 87-110 in T. N. Clark (ed.) Citizen Preferences and Urban Public Policy. Beverly Hills, CA: Sage.

MORLOCK, L. (1974) "Business interests, countervailing groups and the balance of influence in 91 cities," pp. 309-328 in W. D. Hawley and F. W. Wirt (eds.) The Search for Community Power. Englewood Cliffs, NJ: Prentice-Hall.

MUSGRAVE, R. (1959) The Theory of Public Finance. New York: McGraw-Hill.

PRZEWORSKI, A. and H. TEUNE (1970) The Logic of Comparative Social Inquiry. New York: John Wiley.

RAKOVE, M. (1975) Don't Make No Waves, Don't Back No Losers: An Insider's Analysis of the Daley Machine. Bloomington: Indiana University Press.

ROSSI, P. and R. CRAIN (1969) "Small n studies of cities?" Proceedings of the American Statistical Association.

——— (1968) "The NORC Permanent Community Sample." Public Opinion Quarterly 32: 261-272.

ROSSI, P., R. A. BERK, and B. K. EIDSON (1974) The Roots of Urban Discontent. New York: John Wiley.

SCHNEIDER, W. and G. SCHELL (1978) "The new Democrats." Public Opinion (November/December): 7-13.

SCHUMAKER, P. D., R. W. GETTER, and T. N. CLARK (1978) Policy Responsiveness and Fiscal Strain in 51 American Communities: A Manual for Studying City Politics Using the NORC Permanent Community Sample. Prepared for the SETUPS Series of the American Political Science Association.

SCOTT, D. (1976) "Measures of citizen evaluation of local government services," pp. 111-128 in T. N. Clark (ed.) Citizen Preferences and Urban Public Policy. Beverly Hills, CA: Sage.

SHAFFER, W. R. and R. E. WEBER (1974) Policy Responsiveness in the American States. Beverly Hills, CA: Sage.

SMITH, R. A. (1979) "Decision making and non-decision making in cities." American Sociological Review 44 (February).

STONECASH, J. (1978) "Local policy analysis and autonomy: on intergovernmental relations and theory specification." Comparative Urban Research 5: 5-23.

SUDMAN, S. and N. M. BRADBURN (1974) Response Effects in Surveys. Chicago: AVC.

TIDEMAN, T. N. (1976) "The capabilities of voting rules in the absence of coalitions," pp. 23-44 in T. N. Clark (ed.) Citizen Preferences and Urban Public Policy. Beverly Hills, CA: Sage.

TURK, H. (1977) Organizations in Modern Life: Cities and Other Large Networks. San Francisco: Jossey-Bass.

ZIMMERMANN, E. (1979) "Interests and control in community decision-making." Ph.D. dissertation, University of Chicago.

4

Spatial Models of Community Leadership

RONALD S. BURT

MY PURPOSE in this chapter is to offer some propositions for comparative urban policy research—propositions based on spatial models of decision makers. The motivating assumption for these models is that actors within a system are distributed in a multidimensional space where the nature of their distribution in the space says something interesting about their system. Given persons and organizations within a community as actors affecting urban policy decisions, there are two traditions of work on spatial models that can complement one another to inform comparative urban research. Political economists have made assumptions drawn from political science and economics about a system's preference, or policy, space to rigorously derive equilibrium election decisions and probable coalitions. This tradition provides explicit hypotheses on coalition formation and the feasibility of a policy decision that would constitute an equilibrium decision in a community. Drawing from political science and sociology, political sociologists have been especially attentive to the processes by which specific community policies evolve. Sociologists in particular have developed methods for capturing a system's stratification space to describe existing and equilibrium social structure in terms of networks of relations among actors in a system. These developments provide a quantitative, empirical basis for comparative research and describe the manner in which the preference

AUTHOR'S NOTE: *The initial draft of this chapter was written during a leave of absence spent in the Department of Sociology, State University of New York at Albany. That draft benefited from discussion during its presentation at the Conference on Comparative Urban Policy Research sponsored by the National Science Foundation and coordinated by Terry Nichols Clark. Preparation of this final draft was facilitated by a grant from the National Science Foundation (SOC79-25728).*

and stratification spaces in a community change over the course of issue resolution. The basic theme to my remarks here is that these two traditions of work on spatial models complement one another; political economy providing hypotheses about equilibrium policies, sociology providing hypotheses about the structure of the preference space in terms of which those equilibrium policies are derived. I begin with spatial models in sociology, move to those in political economy, and close with propositions taken from the two traditions when considered together.

SPATIAL MODELS IN SOCIOLOGY: NETWORK STRUCTURE IN ISSUE RESOLUTION

Describing the manner in which a system of actors is differentiated by social criteria is an ongoing concern among sociologists. This differentiation is increasingly being analyzed in terms of social networks defining distances within a stratification space.

STRATIFICATION SPACE

The stratification space for a system of N actors is defined in terms of one or more networks of relations among the actors. The relation from actor i to actor j within some network, z_{ij}, has a form given by the strength, or intensity, of i's tendency to interact with j within the network and a content given by the substance of the interaction (for example, social exchange, economic exchange, and so on). A network then consists of all N^2 possible z_{ij} among the N actors in a system. Particularly relevant here are relations in an issue network, the network of relations created among participants in a community conflict over some policy issue. Actors i and j are connected by a strong z_{ij} in an issue network at some point in time to the extent that they have frequent interaction with one another at that time over the issue. The relation is greater than zero to the extent that they share the same attitude toward the issue and negative to the extent that they have opposing attitudes toward the issue. Issue networks can be estimated for large numbers of communities over the course of conflict escalation using ersatz relational data.[1] More traditional relational data could also be obtained, of course, for case studies of individual communities (see Burt, 1980b: 81-89, for a review of such data). Laumann and Marsden (1979) provide a detailed discussion of such data in terms of community oppositional structures.

Given data on relations, the position actor j occupies within an issue network is given by the pattern of relations connecting him to other actors: N relations from each actor i to him and N relations from him to each actor i. In other words, his position is a relational pattern; a vector Z_j containing 2N relations

$$(Z_j = z_{1j}, z_{2j}, \ldots z_{Nj}, z_{j1}, z_{j2}, \ldots z_{jN}).$$

Actors i and j are separated by high distance within the stratification space defined by the issue network to the extent that they occupy very different positions in the network—that is, to the extent that they have very different patterns of relations with other actors. This concept of distance can be analyzed as the Euclidean distance between their respective relations with each actor in the system:

$$d_{ij} = d_{ji} = \left[(Z_j - Z_i)\,(Z_j - Z_i)' \right]^{1/2} = \left[\Sigma_q^N (z_{qj} - z_{qi})^2 + \Sigma_q^N (z_{jq} - z_{iq})^2 \right]^{1/2}.$$

Actors are distributed in the 2N dimensional stratification space defined by these distances according to the similarity of their relational patterns in the issue network. Actors i and j are separated by zero distance when they have identical relations with each actor in the system. They are separated by high distance so as to be on opposite sides of the stratification space to the extent that they have very different relations with each actor in the system. A set of actors separated by zero distances has been discussed as structurally equivalent in recent network models, and the concept of structural equivalence is used to give mathematical/empirical meaning to the classic status/role-set duality. This work has been developed in terms of discrete distance models by White and his colleagues (White et al., 1976; Boorman and White, 1976; Breiger, 1979) and in terms of continuous distance models by myself (1976, 1977b, 1977c). A systematic review of this work and related lines extending to multiple networks is given elsewhere (Burt, 1980b: 100-107, 119-125; 1982: ch. 2).

Given the relational patterns defining a stratification space, sociologists have developed network models predicting actor perceptions and ability to take action. Some of these predictions are reviewed in terms of alternative mathematical models elsewhere (Burt, 1977a, 1980a, 1980c, 1982), and there are some equilibrium models of network structure which explicitly borrow the deductive methods of political economy assuming rational actors (Boorman, 1975; Marsden and Laumann, 1977; Winship, 1977; Burt, 1979b; Marsden, 1981). However, spatial models in sociology tend to be descriptive rather than deductive. They offer parameters for describing social differentiation within the stratification space of a system. Still, the substantive richness of spatial models in sociology complements the deductive power of those in political economy by demonstrating three things connecting actor preferences to a system's stratification space: (1) There is an intimate connection between dense points in a system's stratification space and dense points in the system's preference, or policy, space. (2) There is a systematic change in these two spaces over the course of issue resolution. (3) The extent of change in these spaces is contingent on existing stratification in a system. I will comment briefly on each of these.

HOMOPHILOUS PREFERENCES AND STRUCTURAL EQUIVALENCE

Two actors have homophilous preferences to the extent that they share the same preference on an issue; both want the same issue outcome. Viewing an actor's preference as a subjective evaluation of, or social attitude toward, the possible outcomes of an issue, there is a connection between homophilous preferences and structural equivalence. In a sentence, structurally equivalent actors can be expected to have homophilous preferences. In other words, actors separated by zero distance in the stratification space defined by an issue network can be expected to share the same preferences on the issue. Communication is one reason. Actors who frequently discuss an issue with one another create a shared social understanding of the issue. The strong ties among these actors typically result in them having similar relational patterns so as to be structurally equivalent—that is, separated by zero distance in a stratification space. More generally, homophilous preferences can be attributed to role-playing among structurally equivalent actors. Even if a set of actors have no direct communication relations with one another, but have in common a single pattern of relations with occupants of some distant network position, they will be structurally equivalent and so expected to have homophilous preferences. This expectation is based on actors role-playing; putting themselves in the positions of their peers so as to evaluate alternative actions *as if* they were their peers. The resulting preferences are evaluations of marginal utility as an occupant of one's position in a system. I have elsewhere presented a mathematical model describing the connection between actor preferences and network position and reviewed empirical evidence for the connection (Burt, 1980c, 1982; chs. 5 and 6). Two important qualifications should be mentioned here. First, the connection between preference homophily and distance in a stratification space need not be linear. The two are connected by a power function, so it is quite possible that two actors would share the same preference on some issues only if the two actors were almost perfectly structurally equivalent. Second, the stratification space for a system is no more than an upper limit on the complexity of actor preferences. The tendency for structurally equivalent actors to have the same preferences does not imply that actors separated by high distance have different preferences.

NETWORK STRUCTURE IN CONFILCT ESCALATION

The connection between actor preferences and distance in the stratification space defined by an issue network simplifies with conflict escalation over the issue. As a structural process, community conflict involves systematic change in two aspects of an issue network; the form of conflict is given by change in the pattern of relations in the issue network, and the content of conflict is given by change in the substance of the communications exchanged by participants in the issue network.[2] At any point in time, the form of a conflict is given by the relations and distances among the N actors relevant to the conflict, and its content consists of the substance of their communications with each other. Issues are the same type to the extent that they have the same form and content; that is, to the extent that they similarly involve actors in a community conflict.[3]

The form of conflict escalation during the resolution of an issue involves mobilization and polarization in the issue network. Subsequent to some precipitating event, neutral actors are "mobilized" into the conflict, with internal coordinating relations among allied actors being developed by partisan leaders who emerge within opposing factions. Opponents are "polarized," in the sense that opposition relations develop between actors allied with the separate factions. Those actors allied with both sides of the conflict are neutralized, not so much in the sense that they are no longer involved in the conflict as that they cannot legitimize one side or the other. I have described these processes of mobilization and polarization in detail for community conflicts over nuclear power facilities (Burt, 1979a) and Coleman (1957) provides illustration from diverse types of conflicts. This mobilization and polarization within the issue network serves to increase the distance between opponents and neutralized actors (thereby "removing" them from the conflict) as the distance among allies is decreased. In other words, the distribution of actors in the stratification space defined by an issue network changes with conflict escalation, such that allies draw close together as dense points in the space and these dense points, as opposing factions, pull apart in the space. In their own way distant from the polarized factions, actors neutralized by their own simultaneous commitment to multiple factions are forced into positions scattered through the empty space between opposing factions.

At the same time that the form of an issue network is changing, its content is changing. The content of the network generalizes. The initial controversy sparking conflict weakens existing relations between opponents, thereby inviting other, preexisting, disagreements to be raised. Further, in their efforts to mobilize neutrals to their cause, opponents have an incentive to expand their original disagreement to include new problems that are perhaps more salient to the neutrals they wish to recruit. These two processes ensure that the specific questions over which opponents disagree will multiply as conflict escalates (see Coleman, 1957: 10, on nuclear power facilities, Burt, 1979a). This multiplication of questions ensures that the content of the issue network generalizes. If for no other reason than simplicity, the many specific problems debated initially are replaced with some generalized "type" of problem. This is a crucial step: a shift in the content of communication from disagreement over one or more specific problems to disagreement over something that does not exist in its own right but rather is an abstraction serving to represent the specific problems. The abstractions themselves become complex and are replaced with still more general abstractions concerning basic values and beliefs. These beliefs, outside the original dispute, are not easily subject to change. They typically are not subject to negotiation.[4] By impugning the validity of each other's beliefs, opposing factions (quite reasonably unwilling to change their beliefs) develop emotional dislike for one another. One's opponent is no longer someone like yourself who has come to a different interpretation of a given set of facts. He is some manner of heretic. The original dispute is now a direct antagonism. It is now that the radical elements in factions can rise as partisan leaders to replace moderate leaders.[5] Still worse, the antagonism between opponents is now self-sustaining. As an abstraction, the disagreement between opponents is subject less to negotiation than to normative role definition. Opponents come to define one another as "types" of individuals differing on "types" of issues. Conflict is now self-sustaining, in the sense that the conflict relations developed over the initial controversy now serve to define opposing groups for one another as stereotypical "types" of individuals. The conflict cannot be resolved without redefining the opponents' perceptions of one another. Unfortunately, these opposition relations polarizing factions and the support relations among allies within each faction serve to protect stereotypical images of faction members even if the stereotypes are in error.[6] The same social processes responsible for structurally equivalent actors sharing the same social attitude toward an issue are now responsible for maintaining their stereotypes of participants in the conflict over the issue. In short, the social aspects of conflict have taken over the original dispute. The content of the conflict, the substance of the disagreement between opponents, is now being determined by mobilization and polarization within the issue network. Quite generally, the conflict's form—taken to its extreme—becomes its content.

This point can be emphasized with a more formal treatment of actor preferences. Of the Q questions at issue in a conflict at some point in time, let x_{jq} be actor j's preference with respect to question q. Let p_q be a weight indicating the abstraction of the question. The more abstract a question is, the greater the number of specific questions it subsumes and accordingly the higher p_q will be. Across all questions at issue in the conflict, actors i and j are distant from one another in the preference space defined by the content of the conflict, to the extent that they prefer different outcomes on the most abstract questions. This can be expressed simply as a weighted Euclidean distance analogous to distances in the stratification space defined above:

$$p_{ij} = p_{ji} = \Sigma_q^Q (x_{jq} - x_{iq})\, 2pq½.$$

Generalization of the content of an issue network entails two systematic changes in this distance; Q increases and the p_q shift. These are not random changes. New questions are raised and subsequently abstracted as preexisting controversies are recalled, as neutrals are mobilized into the conflict, and as polarization cultivates antagonism between opposing factions. Thus, the changes in Q and p_q result in increased distance between actors allied with opposing factions and a decrease in the relative distance among actors within each faction. In other words, the distribution of actors in a preference space defined by the p_{ij} changes with conflict escalation in a manner identical to the changes in the stratification space defined by the issue network. This is not too surprising, since the preference space is defined by the content of the issue network and the stratification space is defined by the content and the form of the issue network. The important point here is that the connection between actor preferences and structural equivalence simplifies as conflict over an issue escalates. The general stratification/preference space reduces to a few dense points far removed from one another in the space; each point representing a faction comprised of actors jointly occupying a position in the issue network and sharing similar preferences on the issue.

EXISTING COMMUNITY STRUCTURE

The structure of a community can be described in terms of tendencies for actors to be involved in events together, and network data describing structure in this sense can be obtained in large numbers of communities (Burt, 1981b). These ongoing interactions among community actors result in the day-to-day decisions that constitute community policy. There are strategies by which leaders can prevent decisions from becoming community issues.[7] Where these strategies are effectively employed, or citizens are apathetic, policy decisions need not generate conflict as a community issue. The existing structure of a community, however, affects the likelihood of a given issue arising in the community by affecting mobilization and polarization over the issue.

Depending largely on Olson's (1965: 5-52) theory of collective action, models of mobilization often argue that existing social relations between individuals are necessary to offset each person's natural tendency to let others mobilize in his place. This premise is particularly well articulated by Oberschall (1973: 102-145) and O'Brien (1975: 7-42). In other words, actors have inertia against being mobilized, and this inertia can be overcome by existing social relations. An individual becomes involved in his friend's disputes. An organization becomes involved in its member's disputes. Coleman (1957: 18-23) reviews evidence showing that social relations among citizens in a community facilitates mobilization, and I have reviewed more recent evidence (Burt, 1979a).

While existing relations facilitate mobilization, they inhibit polarization. An individual is unlikely to support a policy if his long-standing friends oppose it. In fact, this is the basis for neutralizing some actors in a conflict. A group or organization that finds half of its membership supporting and half opposing a policy will be neutral in the conflict. This can be stated in stronger terms. Not only does the existence of relations between potential opponents have a negative effect on polarization, the absence of such ties (or worse, the existence of long-standing antagonism) facilitates polarization. Two factions that are strangers or have been in opposition previously, in comparison to two long-standing friends, are likely to become polarized in a conflict that pits them against one another.[8] Here again, Coleman (1957) provides illustrative case study material, and I have reviewed more recent evidence (Burt, 1979a). In particular, interesting analyses of polarization in what I am discussing as a preference space and existing cleavages in community structure are presented by Laumann and Pappi (1976: 163-181) and Laumann and Marsden (1979).

Given knowledge of which actors are likely to support a policy and which are likely to oppose it, in other words, knowledge of the existing structure of a community allows prediction of the probable mobilization and polarization the policy could generate as a community issue. The policy's issue network is nested in the existing community structure.[9] When opponents and supporters of a policy are already segregated in a community's stratification space, a space defined by existing relations among community actors, then the policy is likely to spark a precipitating event, transforming it into an issue-generating community conflict.[10] When supporters and opponents are not segregated in the existing structure, on the other hand, then actors on both sides of the issue are connected by ties dampening polarization over the policy.[11]

SPATIAL MODELS IN POLITICAL ECONOMY: COALITIONS AND EQUILIBRIUM POLICIES

Spatial models in political economy do not contradict those in sociology so much as they involve a different emphasis and methodology. Spatial models in sociology focus on social structure to provide inferences about actor preferences and action given relational patterns defining actor positions in a stratification space. Spatial models in political economy focus on actor preferences to provide hypotheses derived from the distribution of actors in the policy space defined by actor preferences.

POLICY SPACE

Spatial models in political economy define distances between actors in a policy space very similar to the p_{ij} I discussed above in terms of a preference space. Borrowing the notation in Riker and Ordeshook (1973: 310), a citizen's most preferred outcome on issue Q is given by x_q. Across all issues being considered, the citizen's most preferred outcome on all Q issues is a vector X, $X = x_1, x_2, \ldots x_q$, and a system of actors is distributed in a Q dimensional policy space according to their preferred outcomes on issues. Typically, there are two or more candidates in the system who offer a set of issue outcomes as a political platform. Candidate j's position in the policy space is given by a vector of issue outcomes θ_j, $\theta_j = \theta_{j1}, \theta_{j2}, \ldots \theta_{jq}$, where θ_{jq} is his promised outcome on issue Q if he is chosen for office. The distance between a citizen and candidate j in the policy space is high to the extent that the candidate offers issue outcomes very different from those the citizen prefers. There is a clear analogy to the p_{ij} described above in which actors i and j are separated by extensive distance in a preference space to the extent that they have very different preferences. However, the functional form mapping differences between the x_q and θ_{jq} in a general policy space need not be the weighted Euclidean distance formula I used above. Given actor preferences defining a policy space in some manner, hypotheses are derived defining equilibrium policies and the formation of coalitions.

EQUILIBRIUM POLICIES AND COALITIONS

Although the use of spatial models in political economy can be traced back to Hotelling's (1929) analysis of market competition and Black's

(1958) analysis of committee voting, the seminal work for the literature pertinent to this discussion is Downs's (1957) analysis of competition between political parties as if they were market competitors in Hotellings model.[12] Among the central theoretical contributions pertinent here are Plott (1967), Davis et al. (1970), Hinich and Ordeshook (1970), Hinich et al. (1973), and DeSwann (1973: 71-119). These developments are reviewed by Riker and Ordeshook (1973, esp. pp. 307-375). Recent contributions have been made by McKelvey (1975) and McKelvey and Ordeshook (1976). Among recent empirical applications of spatial models in political economy are Aldrich (1975, 1977), Aldrich and McKelvey (1977), Rosenthal and Sen (1977), Chamberlin and Cohen (1978), Fiorina and Plott (1978), and Hinich (1978).

This literature attacks a central problem: What is the "optimum" or "equilibrium" policy for a system of actors, such that a candidate offering this policy would have no incentive to change the policy in two senses? (1) The optimum policy guarantees a minimum plurality not exceeded by adopting any other policy under existing conditions. (2) Changing from the optimum policy lowers or leaves unchanged the candidate's plurality (see Riker and Ordeshook, 1973: 338). An optimum policy would be a Condorcet winner, in the sense that a candidate adopting it would be able to defeat any other candidate's policy in a pairwise comparison of candidates.

An equilibrium policy exists for a system of actors under certain distributions of actors in the system's policy space. It exists if the distribution of actors in the policy space is symmetric and unimodal or bimodal.[13] In other words, if there are one or two dense points in a system's policy space and on each issue, the actors are symmetrically distributed around one or two preferred issue outcomes, then there is an equilibrium policy for the system. Moreover, if each actor's desired outcomes are considered in adopting a policy (that is, if every citizen votes in electing the adopted candidate), then the equilibrium policy is the multivariate mean issue outcome (mean across all actors).[14] As Riker and Ordeshook (1973: 348) point out, this is not a surprising conclusion for a system with a unimodal distribution. If actors are distributed symmetrically around a single point in the policy space, that point seems an obvious equilibrium policy. But the average policy preference can be an equilibrium policy when some actors are alienated by the policy and the distribution of actors is asymmetric (McKelvey and Ordeshook, 1976), or if candidate policy platforms contain ambiguity (McKelvey, 1980). Even if there are two separate points around which actors are distributed, the equilibrium policy is the mean preferred issue outcome rather than the preferred outcome of either point. This same conclusion holds if some actors can be ignored as indifferent to the policy (McKelvey, 1975). However, if actors in any point would be strongly alienated by the mean or the median policy decision—as could be expected if conflict over an issue has escalated to a high intensity—then there is no single equilibrium point in a policy space that is not unimodal. In this circumstance, candidates might adopt policies specific to each dense point (faction) in the space rather than adopt the system multivariate mean or median policy (see Downs, 1957: 119-121; Hinich and Ordeshook, 1970; Riker and Ordeshook, 1973: 348; McKelvey, 1975).[15]

In this absence of a clear equilibrium policy, the presence of multiple factions in the policy space suggests that coalitions will form. Empirical research testing alternative theories of coalition formation find strong support for "policy distance theory" (Leiserson, 1968; Axelrod, 1970; DeSwann, 1973). As stated by DeSwann (1973: 80-119), coalitions formed under policy distance theory need not be minimal in the sense of excluding all but the few actors necessary to determine the adopted policy. Instead, "coalitions emerge from the interaction among actors, each of whom strives to bring about and join a coalition that he expects to adopt a policy which is as close as possible to his own most preferred policy" (p. 82). Building from this spatial basis for coalitions, Riker and Ordeshook (1973: 370) note that as factionalism increases within the policy space, the number of possible winning coalitions increases. The more that resources required to determine the adopted policy are scattered among groups of actors, then the more different combinations of groups there are capable of dominating each other group individually. They go on to note that the probability of any one coalition being a winning coalition over any alternative coalition decreases as this factionalism increases. With increasing numbers of factions in the policy space, there is a probable decrease in the stability of any coalition imposing a policy decision.

COMMENTS

I have now come full circle, beginning with structurally equivalent actors jointly occupying a point in a stratification space and ending with members of a factional coalition defined as a jointly occupied point in a policy space. It seems somewhat futile to attempt to summarize further what has already been a highly condensed discussion. Still, the two traditions of work I have briefly discussed have the potential to inform one another in providing guides for comparative urban policy research. I have emphasized the similarity between spatial models in sociology which focus on social structure as the basis for actor preferences and behavior versus spatial models in political economy which focus on actor preferences as the basis for equilibrium policies and coalitions. Each tradition works toward the other; one starting from actor preferences and making inferences about social relations in factional groups such as coalitions, the other starting from social relations more generally and making inferences about actor preferences. Conflict escalation over an issue makes the similarities between these two traditions much more explicit. Figure 4.1 presents the changes I have described as conflict escalation drawing on both traditions of work on spatial models. The figure corresponds to the "throughput" aspect of the typical input/throughput/output models of comparative urban policy research (see Clark, 1968: 18, 1973: 3, 1975b; Laumann and Pappi, 1976: 14). Times moves from left to right in the diagram. Generally, a policy moves through a preissue stage of activities before some precipitating event transforms it into a community issue-generating conflict. Conflict escalation over the issue proceeds to the extent that the typical proponents and opponents of the policy are a priori segregated in the community's existing structure. Even at this high level of abstraction, it is clear that the history of a policy within a community must be taken into consideration when comparing decisions on the policy across communities. Ostensible differences in two cities' adoption of a policy could be a result of observing them at different stages of conflict escalation. It is possible, however, to suggest some general propositions for comparative research.

To the extent that a policy's proponents are autonomous in their community, the policy will be adopted in a preissue stage of issue resolution. If a policy's supporters can act without constraint from other community actors, they would be expected to implement their preferred policy as a routine decision, ceteris paribus. Given network data on relations among community actors, the structural autonomy of a policy's typical supporters within a city can be determined (Burt, 1980a, 1982: ch. 7). The most important condition to be held constant in assessing this proposition across communities would be leader knowledge of the conflict a policy has generated in other communities. For example, city managers tended to favor water fluoridation in response to its endorsement by the medical community and its low cost; however, the fact that adoption generated severe conflict in other communities led many city managers to hold referenda to see if their own cities would erupt in conflict over the policy (see Crain et al., 1969).

To the extent that significant proponents and opponents of a policy are distributed symmetrically around a multivariate median of their preferences on the policy, a multivariate median version of the policy will be adopted in the community as an equilibrium policy. This adoption could occur during a preissue stage if community leaders are careful to determine citizen preferences so as to tailor the adopted policy to those preferences (see

Preissue Activities		Mobilization, polarization, and generalization in an issue network to the extent that the policy's typical supporters and opponents are a priori segregated in community social structure

TIME & ESCALATING CONFLICT OVER THE POLICY →

Autonomous supporters of the policy adopt it	and/or	An equilibrium policy is adopted as the multivariate median of citizen preferences	and/or	Coalitions form among powerful community actors so as to impose their preferred policy outcome	and/or	The policy is ineffectively implemented to the extent that a coalition-imposed policy is adopted in the context of multiple, polarized factions

Figure 4.1 Predictable Activities by Community Actors as Conflict Escalates over the Adoption of a Community Policy

footnote 7). Alternatively, a compromise version of the policy could be adopted after conflict has begun to escalate. For example, Banfield (1961: 15-56) describes the manner in which a proposed branch hospital was compromised away in an effort to please contending factions. This is not to suggest that an equilibrium policy can be negotiated at any point in time in Figure 4.1. With polarization, the probability of some actors being alienated from a compromise policy increases and the probability of the multivariate median being an equilibrium policy becomes uncertain. Only a winning coalition will be able to impose some version of the policy on the community. In other words, *the longer a policy generates conflict in a community, the less likely it is that any policy outcome will be an equilibrium and the more likely it is that adoption of any version of the policy will require a coalition of actors.* At the same time, conflict escalation and the formation of coalitions are affected by the existing structure of a community so that *conflict can be expected to escalate and encourage the formation of coalitions among powerful, structurally equivalent actors in a community to the extent that a policy's supporters and opponents are a priori segregated in the community's social structure.* Again, network data on relations among community actors are sufficient to make inferences about each actor's prestige or power as control over valued resources (see Burt, 1977a, 1979b, 1982: ch. 2). Structurally equivalent actors would be expected to share similar policy preferences, and those in control of valued community resources would be in a position to form a winning coalition to ensure that their preferred policy outcome was adopted.

However, adopting a policy and implementing a policy are not the same thing. *A coalition-imposed policy decision will be implemented ineffectively to the extent that the decision is far from a multivariate median in the policy space and there are many factions in the policy space.* Ineffective implementation here refers to slow cooperation among the multiple actors required to implement an urban policy and a short life span for the policy within the community (see Sayre and Kaufman, 1960; Levy et al., 1974, on "implementation"). To the extent that there are factions dissatisfied with a policy decision, there is a low probability of cooperation among these factions in effectively implementing the policy decision. Given the increasing probability of a coalition-imposed policy—if any policy is adopted—as conflict escalates, it seems likely that *those communities which become most embroiled in conflict over a policy are least likely to implement the policy effectively.*

NOTES

1. Ersatz network data are interpreted at length elsewhere (Burt, 1981a, 1981b). They are based on interviews with community informants and so are much less expensive than

traditional sociometric-type network data. Despite their low cost, however, they have been demonstrated to provide reliable, valid information on community structure at an aggregate level using data on the Permanent Community Sample (Burt, 1981b). While they can be used to describe intracommunity structure (Burt et al., 1980), their principal advantage lies in providing comparable network data on large numbers of communities.

2. Applying metaphors articulated by Simmel, Coleman (1957: 9-14) discusses these two aspects of community conflict as changes in the social organization of a community and changes in the issues at question in the conflict. The innovative analysis of published case studies by Davis and Weinbaum (1969) is closely related to the idea of an issue network. They use a content analysis of case studies to trace issue networks over time in metropolitan decisions. Given the difficulty of assessing reliability in secondary analyses, however, it is difficult to determine the strongest features of their analysis (see Nelson, 1974).

3. A similar sentiment is reflected in Laumann and Pappi's (1976: 163-172) analysis of "types" of community issues in terms of the different functional sectors of the community activated in support and opposition on each type of issue. Issues are more often classified, however, in terms of a single continuum rather than a whole network. Laumann et al. (1977) discuss issues in terms of their negotiability, instrumental issues being negotiable in contrast to consummatory issues, which are based on nonnegotiable cultural values. A more popular continuum concerns the proportion of a community affected by a policy, areal policies being those which affect the overall community and segmental policies being more special interest policies (Wilson and Banfield, 1964, 1971; Froman, 1967; Clark, 1975a; Smith, 1975).

4. This point has relevance for the idea of classifying community issues in terms of their negotiability (Laumann et al., 1977). To some extent, but by no means completely, negotiability is a function of the time a community spends on an issue. The longer the period during which an issue is generating conflict, then the less negotiable its resolution. There is some evidence in social psychology suggesting that the mere passage of time in the discussion of an issue is capable of polarizing opinion on the issue (Myers and Lamm, 1975; Tesser and Conlee, 1975).

5. Brinton (1965: 148-175) remains a classic discussion of the ouster of moderate leaders by radical leaders. Coleman (1957: 14) discusses this process as "Gresham's Law of Conflict" under which "the harmful and dangerous elements drive out those which would keep the conflict within bounds." Gamson (1975) specifically addresses the strategies of protest groups who feel that they are being excluded from a "meaningful" input into the political process. Both sides in such protests seem unusually willing to use violence in achieving their ends. As put by Simmel (1955: 39) "the parties' consciousness of being mere representatives of supra-individual claims, of fighting not for themselves but only for a cause, can give the conflict a radicalism and mercilessness which find their analogy in the general behavior of certain very selfless and very idealistically inclined persons. Because they have no consideration for themselves, they have none for others either; they are convinced that they are entitled to make anybody a victim of the idea for which they sacrifice themselves."

6. The classic statements here are by Sherif (1935) and Festinger et al. (1950). More recent arguments are provided by Luttbeg and Zeigler (1966), Finifter (1974), Ross et al. (1975), and Stein (1976). Related to this point is the close association Laumann and his colleagues report (Laumann and Pappi, 1976: 152-159; Laumann et al., 1977: 618-622) between values and relations among community elites. The similarity between the values two elites have has a strong direct effect on the tendency for the two elites to discuss community affairs with one another.

7. For example, Meltsner (1971) describes three strategies officials use in what he terms "public avoidance": consensual, where officials perceive a consensus on a policy so that the public can be ignored; negotiatory, where officials discuss proposals with cooper-

ative citizens on a face-to-face basis before implementing decisions; and anticipatory, where officials anticipate public reaction so as to avoid having to negotiate. Further examples are provided by Bachrach and Baratz (1970: 67-103), who describe strategies city officials employed to prevent race riots from occurring in Baltimore in the early 1960s, and Crenson (1971), who describes strategies employed by the steel industry to prevent air pollution from becoming an issue in Gary, Indiana during the 1950s and 1960s. More general discussion of policy decisions which never become community issues is in Wolfinger (1971).

8. A caveat here is that the most intense polarization occurs between erstwhile friends who feel that one has betrayed the other. This caveat, long recognized in metaphor (Simmel, 1955), is also found in empirical experimental work (Swingle, 1966; Swingle and Gillis, 1968).

9. This idea of a community conflict being nested in existing social structure has been discussed in terms of "institutionalization" of exchange (Clark, 1972, 1968: 40-44), the "embedded" quality of exchange relations (Marsden and Laumann, 1977), and the use of exchange relations as "resource surrogates" (Burt, 1977a: 25-36).

10. As Walton proposed (1969: 76): "We refer to these stimuli as triggering events, and propose that they can have their effect either by increasing the magnitude or salience of the issues in the conflict or by lowering one of the barriers to action." Analyzing community mobilization over air pollution in Illinois, for example, Bridgeland and Sofranko (1975) report that the best predictor of mobilization is not the actual level of pollution but rather the occurrence and publication of some event demonstrating the potentially dangerous consequences of pollution (such as property damage, livestock or fish kills, evacuations, of hospitalizations). When controlling for the occurrence of such a precipitating event, the actual level of pollution in a community had no association with mobilization to combat pollution.

11. Kapferer (1969) offers a detailed description of the manner in which interpersonal ties limit social conflict. My generalization here assumes that the policy being considered is not capable of rupturing the existing social ties between supporters and opponents. An issue rupturing existing ties would be expected to generate the most severe levels of polarization (see footnote 8).

12. Two related lines of work in political economy also trace their ancestry to Downs's predecessors. Refined by Smithies (1941), Hotelling's spatial analysis of market competition is the forerunner of location theory in economics (see Beckmann, 1968, for an introduction). Elegantly stated in Arrow's (1951) impossibility theorem, Black's analysis of voting rules is the basis for social choice theory in political economy (see Riker and Ordeshook, 1973: 78-115, for an introduction, or Sen, 1970, for a more thorough introduction).

13. I am ignoring the question of how actors perceive distance between their preferred policy outcome and the outcome promised as a political platform. This question is addressed in the political economy literature under assumptions about citizen utility functions (see Riker and Ordeshook, 1973: 312-322). The basic question is: How does an actor's utility with having θ_j adopted as policy decrease with increasing dissimilarity between θ_j and his desired outcomes X_j? If equilibrium can be reached under any utility function assumed in the literature, I am treating it as possible for the purposes of this discussion. There is little empirical support for one utility function over its alternatives.

14. This important result is due to Plott (1967) and appears to have unanticipated empirical support in comparison to coalition models (Fiorina and Plott, 1978).

15. A possible exception here is the unlikely situation in which a candidate can generate "extremist" support for the median policy from extremists who are symmetrically distributed around the median policy beyond the bimodal points on either side of the medial policy (McKelvey, 1975; Aldrich, 1977). Another route to determining an equilibrium policy in the context of multiple factions in the policy space is through a series of votes; each

issue outcome proposed under a policy must obtain a majority over each corresponding issue outcome under other policies (Kramer, 1972).

REFERENCES

ALDRICH, J. (1977) "Electoral choice in 1972: a test of some theorems of the spatial model of electoral competition." Journal of Mathematical Sociology 5: 215-237.

——— (1975) "Candidate support functions in the 1968 election: an empirical application of the spatial model." Public Choice 22: 1-22.

——— and R. McKELVEY (1977) "A method of scaling with applications to the 1968 and 1972 presidential elections." American Political Science Review 71: 111-130.

ARROW, K. J. (1951) Social Choice and Individual Values. New Haven: Yale University Press.

AXELROD, R. (1970) Conflict of Interest: A Theory of Divergent Goals with Applications to Politics. Chicago: Markham.

BACHRACH, P. and M. S. BARATZ (1970) Power and Poverty: Theory and Practice. New York: Oxford University Press.

BANFIELD, E. C. (1961) Political Influence. New York: Free Press.

BECKMANN, M. (1968) Location Theory. New York: Random House.

BLACK, D. (1958) The Theory of Committees and Elections. Cambridge: Cambridge University Press.

BOORMAN, S. A. (1975) "A combinatorial optimization model for transmission of job information through contact networks." Bell Journal of Economics 6: 216-249.

——— and H. C. White (1976) "Social structure from multiple networks. II. role structures." American Journal of Sociology 81: 1384-1446.

BREIGER, R. L. (1979) "Toward an operational theory of community elite structures." Quality and Quantity 13: 21-57.

BRIDGELAND, W. M. and A. J. SOFRANKO (1975) "Community structure and issue-specific influences: community mobilization over environmental quality." Urban Affairs Quarterly 11: 186-214.

BRINTON, C. (1965) The Anatomy of Revolution. New York: Vintage.

BURT, R. S. (1982) Toward a Structural Theory of Action: Network Models of Social Structure, Perception and Action in a System of Actors. New York: Academic Press.

——— (1981a) "Studying status/role-sets as ersatz network positions in mass surveys." Sociological Methods & Research 9: 313-337.

——— (1981b) "Comparative power structures in American communities." Social Science Research 10: 115-176.

——— (1980a) "Autonomy in a social topology." American Journal of Sociology 85: 892-925.

——— (1980b) "Models of network structure." Annual Review of Sociology 6: 79-141.

——— (1980c) "Actor interests in a social topology: foundation for a structural theory of action." Sociological Inquiry 49: 107-132.

——— (1979a) "Community conflict in the nuclear power issue: a review exploring the network structure of conflict escalation." Working Paper 6. Berkeley: Survey Research Center, University of California.

——— (1979b) "Relational equilibrium in a social topology." Journal of Mathematical Sociology 6: 211-252.

——— (1977a) "Power in a social topology." Social Science Research 6: 1-83.

——— (1977b) "Positions in multiple network systems, part one: a general conception of stratification and prestige in a system of actors cast as a social topology." Social Forces 56: 106-131.

——— (1977c) "Positions in multiple network systems, part two: stratification and prestige among elite decision-makers in the community of Altneustadt." Social Forces 56: 551-575.

——— (1976) "Positions in networks." Social Forces 55: 93-122.

BURT, R. S., K. L. LIEBEN, and M. G. FISCHER (1980) "Network power structures from informant perceptions." Human Organization 39: 121-133.

CHAMBERLIN, J. R. and M. D. COHEN (1978) "Toward applicable social choice theory: a comparison of social choice functions under spatial assumptions." American Political Science Review 72: 1341-1356.

CLARK, T. N. (1975a) "The Irish ethic and the spirit of patronage." Ethnicity 2: 305-359.

——— (1975b) "Community power," pp. 271-295 in Annual Reviews of Sociology, Vol. 1. Palo Alto, CA: Annual Reviews, Inc.

——— (1973) Community Power and Policy Outputs. Beverly Hills, CA: Sage.

——— (1972) "Structural-functionalism, exchange theory, and the new political economy: institutionalization as a theoretical linkage." Sociological Inquiry 42: 275-298.

——— [ed.] (1968) Community Structure and Decision-Making: Comparative Analysis. San Francisco: Chandler.

COLEMAN, J. S. (1957) Community Conflict. New York: Free Press.

CRAIN, R. L., E. KATZ, and D. B. ROSENTHAL (1969) The Politics of Community Conflict: The Fluoridation Decision. Indianapolis: Bobbs-Merrill.

CRENSON, M. A. (1971) The Un-Politics of Air Pollution. Baltimore: Johns Hopkins University Press.

DAVIS, M. and M. G. WEINBAUM (1969) Metropolitan Decision Processes: An Examination of Case Studies. Chicago: Rand McNally.

DAVIS, O., M. HINICH, and P. ORDESHOOK (1970) "An expository development of a mathematical model of the election process." American Political Science Review 64: 426-448.

DeSWANN, A. (1973) Coalition Theories and Cabinet Formations. San Francisco: Jossey-Bass.

DOWNS, A. (1957) An Economic Theory of Democracy. New York: Harper & Row.

FESTINGER, L., S. SCHACTER, and K. BACK (1950) Social Pressures in Informal Groups. Stanford: Stanford University Press.

FINIFTER, A. W. (1974) "The friendship group as a protective environment for political deviants." American Political Science Review 68.

FIORINA, M. P. and C. R. PLOTT (1978) "Committee decisions under majority rule: an experimental study." American Political Science Review 72: 575-595.

FROMAN, L. A., Jr. (1967) "An analysis of public policies in cities." Journal of Politics 29: 94-108.

GAMSON, W. A. (1975) The Strategy of Social Protest. Homewood, IL: Dorsey Press.

HINICH, M. J. (1978) "Some evidence on non-voting models in the spatial theory of electoral competition." Public Choice 33: 83-102.

——— and P. C. ORDESHOOK (1970) "Plurality maximization vs. vote maximization: a spatial analysis with variable participation." American Political Science Review 64: 772-791.

HINICH, M. J., J. O. LEDYARD, and P. C. ORDESHOOK (1973) "A theory of electoral equilibrium: a spatial analysis based on the theory of games." Journal of Politics 35: 154-193.

HOTELLING, H. (1929) "Stability in competition." Economic Journal 39: 41-57.

KAPFERER, B. (1969) "Norms and the manipulation of relationships in a work context," pp. 181-244 in J. C. Mitchell (ed.) Social Networks in Urban Situations. Manchester, England: Manchester University Press.

KRAMMER, G. (1972) "Sophisticated voting over multidimensional choice spaces." Journal of Mathematical Sociology 2: 165-180.

LAUMANN, E. O. and P. V. MARSDEN (1979) "The analysis of opposition structures in political elites: identifying collective actors." American Sociological Review 44: 713-732.

LAUMANN, E. O. and F. U. PAPPI (1976) Networks of Collective Action: A Perspective on Community Influence Systems. New York: Academic Press.

——— and J. GALASKIEWICZ (1977) "Community elite influence structures: extension of a network approach." American Journal of Sociology 83: 594-631.

LEISERSON, M. (1968) "Factions and coalitions in one-party Japan: an interpretation based on the theory of games." American Political Science Review 57: 770-787.

LEVY, F., A. MELTSNER and A. WILDAVSKY (1974) Urban Outcomes: Schools, Streets, Libraries. Berkeley: University of California Press.

LUTTBEG, N. A. and H. ZEIGLER (1966) "Attitude consensus and conflict in an interest group: an assessment of cohesion." American Political Science Review 60: 655-666.

McKELVEY, R. (1980) "Ambiguity in spatial models of policy formation." Public Choice 35: 385-402.

——— (1975) "Policy related voting and electoral equilibria." Econometrica 43: 815-843.

——— and P. C. ORDESHOOK (1976) "Symmetric spatial games without majority rule equilibria." American Political Science Review 70: 1172-1184.

MARSDEN, P. V. (1981) "Introducing influence processes into a system of collective decisions." American Journal of Sociology 86.

——— and E. O. LAUMANN (1977) "Collective action in a community elite: exchange, influence resources and issue resolution," pp. 199-250 in R. J. Liebert and A. W. Imershein (eds.) Power, Paradigms, and Community Research. Beverly Hills, CA: Sage.

MELTSNER, A. J. (1971) The Politics of City Revenue. Berkeley: University of California Press.

MYERS, D. G. and H. LAMM (1975) "The polarizing effect of group discussion." American Scientist 63: 297-303.

NELSON, M. D. (1974) "The validity of secondary analyses of community power studies." Social Forces 52: 531-537.

OBERSCHALL, A. (1973) Social Conflict and Social Movements. Englewood Cliffs, NJ: Prentice-Hall.

O'BRIEN, J. (1975) Neighborhood Organization and Interest Group Processes. Princeton: Princeton University Press.

OLSON, M. (1965) The Logic of Collective Action. Cambridge: Harvard University Press.

PLOTT, C. R. (1967) "A notion of equilibrium and its possibility under majority rule." American Economic Review 57: 787-806.

RIKER, W. H. and P. C. ORDESHOOK (1973) An Introduction to Positive Political Theory. Englewood Cliffs, NJ: Prentice-Hall.

ROSENTHAL, H. and S. SEN (1977) "Spatial voting models for the French Fifth Republic." American Political Science Review 71: 1447-1466.

ROSS, L., M. R. LEPPER, and M. HUBBARD (1975) "Perseverance in self-perception and social perception: biased attributional processes in the debriefing paradigm." Journal of Personality and Social Psychology 32.

SAYRE, W. S. and H. KAUFMAN (1960) Governing New York City: Politics in the Metropolis. New York: Russell Sage.

SEN, S. (1970) Collective Choice and Social Welfare. San Francisco: Holden Day.

SHERIF, M. (1935) A Study of Some Social Factors in Perception. Archives of Psychology, Vol. 187.

SIMMEL, G. (1955) Conflict and the Web of Group-Affiliations (1908; K. H. Wolff and R. Bendix, trans.). New York: Free Press.

SMITH, R. A. (1975) "Community power and decision-making: a replication and extension of Hawley." American Sociological Review 41: 691-705.

SMITHIES, A. (1941) "Optimum location in spatial competition." Journal of Political Economy 69: 423-439.

STEIN, A. A. (1976) "Conflict and cohesion: a review of the literature." Journal of Conflict Resolution 20: 143-172.

SWINGLE, P. G. (1966) "Effects of the emotional relationships between protagonists in a two-person game." Journal of Personality and Social Psychology 4: 270-279.

——— and G. GILLIS (1968) "Effects of the emotional relationship between protagonists in the prisoner's dilemma." Journal of Personality and Social Psychology 8: 160-165.

TESSER, A. and M. C. CONLEE (1975) "Some effects of time and thought on attitude polarization." Journal of Personality and Social Psychology 31: 262-270.

WALTON, R. E. (1969) Interpersonal Peacemaking: Confrontations and Third Party Consultation. Reading, MA: Addison-Wesley.

WHITE, H. C., S. A. BOORMAN, and R. L. BREIGER (1976) "Social structure from multiple networks. I. blockmodels of roles and positions." American Journal of Sociology 81: 370-380.

WILSON, J. Q. and E. C. BANFIELD (1971) "Political ethos revisited." American Political Science Review 65: 1048-1062.

——— (1964) "Public regardingness: a value premise in voting behavior." American Political Science Review 58: 876-887.

WINSHIP, C. (1977) "The allocation of time among individuals," pp. 75-100 in K. F. Schuessler (ed.) Sociological Methodology 1978. San Francisco: Jossey-Bass.

WOLFINGER, R. E. (1971) "Nondecisions and the study of local politics." American Political Science Review 65: 1063-1080.

5

Investments in Local Infrastructure

LARRY E. HUCKINS
GEORGE S. TOLLEY

URBAN INFRASTRUCTURE—schools, streets, sewers, water supply facilities, fire stations and the like—provides valuable services demanded by the households and businesses of the nation's cities. Most fundamentally, urban infrastructure is part of the larger capital stock of the nation. Elsewhere (Tolley and Huckins, forthcoming) we have estimated that municipal urban infrastructure may account for as much as five percent of the nation's capital stock. Like other capital, urban infrastructure is long-lived, but it wears out eventually. Providing urban infrastructure requires a number of complex decisions pertaining to capital services, maintenance, depreciation, financing, and taxation.

Attention to urban infrastructure is increasing, largely because of the financial problems and signs of economic decline of some of the larger cities. The economics of urban infrastructure has been relatively neglected. The real capital assets component of local government balance sheets (by far a neglected accounting concept in the public sector) is the largest and dwarfs the magnitude of city financial assets. Capital assets also typically exceed liabilities, including short- and long-term debt outstanding, unfunded pensions, and long-term leasing obligations of cities, although little work as been done in this area to date (however, see Tolley and Huckins, forthcoming: ch. 7). In this chapter we consider concepts needed to understand and evaluate questions being raised about urban infrastructure. A question being asked frequently is whether urban infrastructure is wearing out and failing to be replaced. An associated question is whether financial problems of cities are leading them to neglect investment and upkeep of infrastructure. Concern with the consequences is heightened by the irreversibility question—that is, whether beyond a point of neglect it will be much more expensive to restore the infrastructure than to keep it up in the first place. Further questions concern how urban infrastructure decisions are related to debt and finance, with the implication that infrastructure and debt problems of cities may be self-reinforcing. Still further questions concern the efficiency with which expenditures are made, as well as the magnitude and nature of the federal aid for capital expenditures on which cities have become increasingly dependent.

In the first section, a stock adjustment approach to understanding municipal capital expenditures is described within which questions related to the wearing out and replacement of infrastructure are discussed. Local governments are decision-making authorities who must choose the combination of inputs to produce services demanded by community residents considering both relative input prices and adjustment costs. The critical interdependency of maintenance expenditures and capital quality is developed in the second part, and the consequences of improper decisions on service outputs are noted.

A LABOR-CAPITAL STOCK ADJUSTMENT MODEL

The focus of most local public expenditure studies has been on particular categories of urban services. Most often the dependent variable has been current expenditures. While it might be tempting to think of taking yearly capital expenditures as a dependent variable and analyzing it in the same way, there would be several shortcomings to this approach. First, municipalities do not demand investment expenditures. They demand a total capital stock from which they obtain services. Second, the services from capital are used in conjunction with services of labor to obtain outputs. The use of capital and labor in conjunction with one another needs to be analyzed, not only to explain long-run substitutabilities but also to explain the joint roles of labor and capital in short-run adjustments. Third, only if investment is related to the capital stock demanded can one understand long-term and short-term patterns of expenditures. Investment is not a direct response to income and price variables, but is rather a response to attain desired or equilibrium capital stocks. When the desired capital stock changes, investment being small in relation to the total stock may change by a great amount.

Municipal governments serve as economic decision units, in that they make choices about the allocation of resources through choosing both the mix of services to be provided and the amounts of factors to be utilized in producing services. Municipal services are produced using capital inputs as well as labor, although to date existing studies of state and local government expenditures have failed to incorporate capital explicitly into their models. Huckins (1980) estimates derived demands for capital and labor building on studies of the demands for governmental services by Borcherding and Deacon (1972), Bergstrom and Goodman (1973), Pommerehne and Frey (1976), and Perkins (1977). Certain problems arise as a result of the fact that municipal governments are different from private firms, especially in their objective functions, in the way they raise revenues, and in the way they price their services. These problems are handled by incorporating a theory of public choice into the analysis.

Municipal governments are initially assumed to be able to purchase inputs (labor and capital) in each period at exogenously given prices. However, local governments incur adjustment costs in changing their stock of capital or the effective size of their labor force. In the model, these adjustment costs rise at an increasing rate as the absolute amount of net investment or aggregate number of manhours employed rises. Adjustment costs occur due to planning and construction time requirements; installation costs; costs of selling, detaching, and moving equipment and structures; training costs; recruitment expenses; and overtime premiums incurred until the number of employees is sufficiently increased. Adjustment costs explain why it may be optimal for municipal governments to deviate from long-run equilibrium levels of capital. Nadiri and Rosen (1973) point out in their study of American manufacturing sectors that in a general model labor will also not be employed at its equilibrium level, even given the stock of capital in each period, if there are costs of adjustment for labor as well.

As in the historical studies of capital formation conducted by Goldsmith (1955) and Kendrick (1976), and also in the studies of private sector capital investments by Jorgenson (1967), Nadiri and Rosen (1973), and Hodge (1979), Huckins assumes that capital depreciates at a constant rate. The rate of depreciation of capital is, of course, subject to variation in response to different streams of maintenance expenditures, different types of capital, different intensities of usage, different climatic conditions, and other factors which vary across cities. In the absence of information on how depreciation rates do vary cross-sectionally or over time in response to these endogenous and exogenous factors, the assumption of a constant rate of depreciation, d, is retained for initial work. Thus, net investment is given by the familiar declining balance formula:

$$dK/dt = I\ (t) - d\ K\ (t) \qquad 0 < d < 1 \qquad [1]$$

Capital stock estimates were generated for each of the 46 cities with populations over 300,000 in 1970 using annual data on capital outlays from the *City Government Finances* series. Initial stocks were estimated to be twice the level of debt outstanding in 1932. A real depreciation rate of three percent was chosen based on weighting engineering estimates of depreciation for major capital types by the average relative importance of the infrastructure type in municipal budgets (alternative rates were also utilized to check the sensitivity of the estimates). Capital outlays were deflated using either the gross national product price deflator or the Department of Commerce composite construction cost index. Stock estimates are biased downward for cities annexing already developed areas. The bias is lower the higher the rate of depreciation and the more distant the annexation was in the past. Any biases are also lower if initial stock estimates are inaccurate the higher the rate of depreciation. Service flows are assumed to be proportional to the stock of capital.

Each municipal government chooses its level of capital and labor to maximize the flow of services to a median voter net of input and adjustment costs subject to equation 1, initial starting stocks of capital and labor, and nonnegativity constraints on labor and capital. Based on the disequilibrium model of factor demands, Huckins estimated a generalized Koyck adjustment model for 46 cities for the period 1955-1977 with capital and labor as dependent variables.

Both capital and labor were found to be related to the demand shifter variables included in the model, as expected. However, capital does not appear to behave as a cyclical shock absorber as one might expect. Labor fluctuates relatively more in response to changes in income than does capital. According to the coefficient on the lagged dependent variables, the capital stock adjusts to long-run equilibrium levels in about 15 years, whereas labor adjustment takes only half as long.

Cities were found to be very responsive to factor price changes in the long run. The long-run elasticity of substitution for the entire sample was estimated to be about 2 in absolute value. Estimates for individual cities ran as high as 5. Capital and labor acted like substitutes in each of the 46 cities—a finding at least consistent with (although not proving) rational cost-minimizing behavior by municipal governments. This finding of high substitutability is important to the evaluation of the long-run impacts of subsidy programs to one factor or the other (for example, labor subsidies or public works subsidies), although impacts appear to take a long time to work out completely given the long adjustment time found.

MAINTENANCE AND IRREVERSIBILITIES

While obtaining capital stock estimates derived from capital outlay series appears to be the only feasible approach given present data availability, the estimates neglect the effect of maintenance on depreciation rates and capital stock condition. Expenditures for maintenance or upgrading the condition of capital which lead to lower costs of providing services flows in future periods are a form of capital, although maintenance expenditures are typically not classified as capital outlays in the available data. Current operation expenditures have a large element of capital expenditure, in that they include expenditures for maintenance which affect the future services available from capital. Maintenance keeps capital from deteriorating in ways that would be more expensive or virtually impossible to reverse in the future.

The capital component of current expenditures varies markedly by type of government expenditure. Current road expenditures are in large part really capital expenditures whose benefits extend over many years in the future. Repair of potholes prevents further deterioration of roadbeds to stages where travel becomes slow and hazardous or damaging to autos, trucks, and cargo. Current education expenditures, on the other hand, are virtually all for wages and salaries of teachers and administrators rendering current services not much affecting the future condition of school buildings. This is because the primary purpose of education is to build up a stock of human capital consisting of knowledge, skills, and the like and not a network of physical infrastructure, as is the case for highway expenditures.

The value of a capital good following an initial investment is a function of the stream of maintenance expenditures over time. The value of a capital asset, K, at time r, which was constructed at time t, may be written

$$K_{r,t} = K(I_t, M_{t+1}, M_{t+2}, \ldots, M_{r-1}, M_r) \qquad [2]$$

The optimal stream of maintenance expenditures will depend on the community's desired level of capital stocks, on the available production technology, and on the opportunity cost of funds which depends on the real discount rate and any liquidity constraints facing the community. The marginal product of maintenance is expected to be a declining function of the level of maintenance in any time period. Changes in the opportunity cost of funds over the business cycle caused by liquidity constraints or revenue shortfalls may lead to the greater cyclicality of maintenance expenditures over time. While expenditures on maintenance are not directly reported in governmental financial statistics, fluctuations in maintenance expenditures might be estimated by netting out from current operations expenditures that part of labor used to produce current services. The amounts of labor and materials used in producing current services could arbitrarily be assumed to be proportional to the capital stock, so that estimates of a residual component—maintenance—could then be made.

An iterative procedure could be followed starting with yearly capital stocks (estimated assuming a constant depreciation rate). Maintenance inputs could be estimated as a residual by subtracting out estimated current service inputs assumed proportional to capital stock estimates. The resulting maintenance series could then be used to adjust the depreciation in each year, arriving at a new set of yearly capital stock estimates, which would provide a starting point for estimating a revised estimated maintenance series. The iteration might continue until each year's estimated capital stock were little affected by further iteration. The new capital stock series could be compared with the initial stock estimates to indicate the possible magnitudes of biases from assuming a constant rate of depreciation. The new stock estimates may also be relevant for investigating the fluctuations in municipal debt over business cycles and during periods of fiscal crises.

A key issue is at what point a decline in maintenance expenditures leads to a pronouncedly severe decline in capital quality in the future. While temporary postponement of maintenance expenditures will result in a slightly lower quality capital stock and higher costs of restoration later on, the severe consequence of major collapse or service failure are likely to occur only after sustained neglect. The chance of major catastrophes, such as the collapse of a main water distribution tunnel or collapse of a bridge or elevated rail or highway structure, is much greater after a long period of neglect than after a short period given deterioration of most structures occurs only gradually. Federal aid has been very important to New York City, enabling it to resuscitate its capital program and prevent the severe consequences of cumulative neglect. Other cities with less dramatic problems have not been targets of such special rescue efforts, although federal aid to cities was substantially increased in the 1970s. Whether cities have neglected types of infrastructure not supported by federal efforts is not known, and the potential for major failures in infrastructure systems is uncertain, since little documentation of infrastructure conditions is available. Even the studies of New York City and Cleveland by the Urban Institute cover only a limited number of infrastructure types, and in most cities detailed surveys have not been conducted.

The effect of maintenance on future capital quality can be expected to be enormously different for different types of infrastructure. Postponement of maintenance expenditures for a given period of time is likely to have a more serious impact on short-lived capital such as equipment. Failure to change the oil and to grease and tune up city police cars, firetrucks, sanitation vehicles, and the like will result in increased frequency of major breakdowns of the city's rolling capital stock. Such neglect in New York City may explain the current rapid deterioration in subway services there. Structures, water and sewer lines, bridges, roads, and other long-lived assets will typically be less severely affected in the short run from postponed maintenance.

The answer to the question of how, in quantitative terms, present maintenance and capital refurbishing expenditures affect the quality of municipal capital stocks at various points of time in the future is likely to be at odds with the more rigid forms of the engineering requirements approach. A dollar less expenditure on maintenance results in less capital quality in the future, not instant and complete collapse. The time path of maintenance expenditures is likely to be important to the condition and longevity of capital. Trying to compensate for neglect or undesirably low maintenance expenditures in one period by making above-normal levels of expenditures in the next period is likely to result in a higher total cost of achieving a given level of capital quality than if normal maintenance expenditures were made in each period. This is because repair and maintenance tend to become increasingly expensive the longer they are postponed. This pattern of expenditure may be justified in circumstances where temporary revenue shortfalls and urgent alternative expenditure requirements in the first period make the extra costs from postponing maintenance worthwhile. Circumstances which would lead cities to make abnormally high maintenance expenditures in the first period in anticipation of neglect in a later period are more difficult to discern. This is because total maintenance expenditures required would likely be greater in order to maintain a certain quality of capital. Cities would be unlikely to choose to forego fulfilling alternative demands when the costs of shifting some funds away from maintenance in the first period in terms of reduced service flows is low.

The capital spending process has inertia, and the consequences of the decisions may be long delayed—in some cases, capital spending is deferable without short-run consequence but with possibly severe and irreversible long-run declines in service levels. As noted earlier, public capital spending varies less than current spending during business cycles (contrary to behavior in the private sector). When a broader definition is utilized which incorporates capital produced from maintenance expenditures, the finding may change. Cities may in practice let their capital stocks be a shock absorber when they face changes in revenue receipts or credit market conditions over the business cycle or in their underlying economic condition.

Detailed knowledge about the depreciation paths for different types of infrastructures is lacking, and little is known about the effects of maintenance on depreciation. Both topics are important for further research. Cities typically have not separated out expenditures for maintenance from expenditures for providing current services. Nor have they kept adequate records over time of maintenance and repair work done to maintain the quality of the capital stock. Engineers have provided estimates of desirable rates of replacement or repair of infrastructure, although their estimates may be unduly pessimistic. New York City engineers said that water mains should be replaced every 100 years. Yet even though many pipes have now exceeded that age, the number of breakages, according to David Grossman has, not increased significantly in recent years as one would expect if New York's aging water system were really wearing out. Due to the lack of evidence about the adequacy of maintenance expenditures, the importance of the irreversibility problem remains moot.

REFERENCES

ASHENFELTER, O. C. and R. C. EHRENBERG (1975) "The demand for labor in the public sector," pp. 55-78 in D. S. Hamermesh (ed.) Labor in the Public and Nonprofit Sectors. Princeton: Princeton University Press.

BERGSTROM, T. C. and R. P. GOODMAN (1973) "Private demands for public goods." American Economic Review 63: 280-296.

BORCHERDING, J. E. and R. T. DEACON (1972) "The demand for the services of non-federal governments." American Economic Review 62: 891-901.

CLARK, T. N. and L. C. FERGUSON (forthcoming) Political Processes and Urban Fiscal Strain. The Twentieth Century Fund.

EISNER, R. and R. S. STROTZ (1963) "Determinants of business investment," pp. 59-337 in D. B. Suits et al. (eds.) Impacts of Monetary Policy. Englewood Cliffs, NJ: Prentice-Hall.

EHRENBERG, R. G. (1972) The Demand for State and Local Government Employees. Lexington: D.C. Heath.

GARDNER, J. L. (1972) "City size and municipal service costs." Urban Economics Report No. 77. University of Chicago.

GOLDSMITH, R. W. (1955) A Study of Saving in the United States. Princeton: Princeton University Press.

HANSEN, N. (1965) "Municipal investment requirements in a growing agglomeration." Land Economics 41: 49-56.

HODGE, J. H. (1979) "A study of regional investment decisions." Ph.D. dissertation, University of Chicago.

HUBBELL, L. (1979) Fiscal Crisis in American Cities: The Federal Response. Cambridge, MA: Ballinger.

HUCKINS, L. E. (1980) "Large municipal government factor demands and factor substitutions." Urban Economics Report. University of Chicago.

——— (1979) "Investments in local public infrastructure." (unpublished)

JORGENSON, D. W. (1967) "The Theory of investment behavior," in R. Ferber (ed.) Determinants of Investment Behavior. New York: National Bureau of Economic Research.

KENDRICK, J. W. (1976) The Formation and Stocks of Total Capital. New York: National Bureau of Economic Research.

NADIRI, M. I. and S. ROSEN (1973) A Disequilibrium Model of Demand for Factors of Production. New York: National Bureau of Economic Research.

PERKINS, G. M. (1977) "The demand for local public goods: elasticities of demand for own price, cross prices, and income." National Tax Journal 30: 411-422.

PETERSEN, J. E. (1968) "The determinants of state and local government capital outlays." Ph.D. dissertation, University of Pennsylvania.

POMMEREHNE, W. W. and B. S. FREY (1976) "Two approaches to estimating public expenditures." Public Finance Quarterly 4: 395-407.

TOLLEY, G. S. and L. E. HUCKINS (forthcoming) Urban Capital and the Future of the Cities. The Twentieth Century Fund.

TOLLEY, G. S., P. E. GRAVES, and J. L. GARDNER (1979) Urban Growth Policy in a Market Economy. New York: Academic Press.

USHER, D. [ed.] (1980) The Measurement of Capital. National Bureau of Economic Research. Chicago: University of Chicago Press.

WEICHER, J. C. and R. J. EMERINE III (1973) "Econometric analysis of state and local aggregate expenditure functions." Public Finance 28: 69-83.

6

The Power of Business in the City

ROGER FRIEDLAND
WILLIAM T. BIELBY

IN THIS CHAPTER we address the question of business power and local policy in U.S. cities, a topic which has occasioned almost continuous academic controversy in the post-World War II period. Political sociologists have viewed business power as potentially dangerous, blocking channels for popular participation. In human ecology, business power has been seen as a benign, if not inevitable, outcome of economic dominance. That power was considered necessary for the local political system to identify needs and implement policies. These conflicting theoretical assumptions have generated diverse methods of measuring business power. The methods in turn have come to identify different attributes of business presumed to affect public policy. Below, we review three such attributes that affect business power over local policy, as well as the theoretical debates over how to interpret them. In the subsequent section we indicate the shortcomings of each of the three methods of assessing business influence.

POLITICAL POWER AND URBAN BUSINESS

Three research traditions have attempted to measure the power of business over urban policy. Each tradition identified a different attribute as the determinant of the political power of urban business. The first tradition focused on patterns of political participation, the second

AUTHORS' NOTE: *An earlier version of this chapter was presented at the Conference on Urban Policy Research, University of Chicago, April 26-27, 1979. James N. Baron provided invaluable research assistance.*

stressed the local organizational structure of business, while the third tradition considered the economic activity and function of local business. Within each research tradition, these attributes were related empirically to urban public policy. These empirical relationships were then used to infer the political power of business.

POLITICAL PARTICIPATION AND BUSINESS POWER

In this first approach, the primary attribute of business, and the source of its political power, is the political participation of businessmen. The political power of business is observed in its successful attempts to shape urban policy. This approach identifies political participants in policy formation by their business role and relates their expressed policy preferences and the volume of their participation to policy outcomes. In this approach, neither the participation of business elites in urban politics nor their influence on urban policy was at issue. The empirical debate centered instead on the extent to which business elites influenced urban policy, the extent to which their influence was tempered by other groups, and the extent to which other kinds of participants were also able to affect urban policy. In short, researchers attempted to measure the relative volume and efficacy of business participation in the formation of local public policies.

Two seemingly conflicting research strategies emerged to measure the volume and efficacy of business participation in urban politics. The reputationalists (Hunter, 1953; Miller, 1958) asked key informants to identify the uncontested leaders on a variety of issues (for example, attracting new industry or city development plans), while the decisionalists (Polsby, 1963; Dahl, 1961) asked informants to identify those who participated and those who prevailed in actual decisions made by the city government. The former researchers tended to find greater business influence—greater volume and less contested participation—than did the latter.

But on closer examination, the similarities between these two kinds of studies are more striking than are the differences in their findings. Both strategies assume that business political participation is at least necessary for business to have political power. Both examine the impact of that participation on the issues which had penetrated the public arena and on the subsequent decisions that were made to resolve those issues. Hypothetical or real issues are sampled and patterns of reputed or real political participation are aggregated across these issues. The extent of business influence in a particular city is inferred from the extent to which participants in political issues are drawn disproportionately from businessmen and the extent to which the positions articulated by businessmen prevail.[1]

The empirical focus of these studies is on *individual* businessmen—whether retail, bank, or industrial owners or managers—and their participation. It is individual businessmen who have preferences and can choose to use their political resources to realize them in the political arena. The assumption of these studies is that it is possible to agregate individual acts of business participation as if they were equal.

The empirical focus of both kinds of studies is on those concerns which had already become public issues. The role of political participation in generating those issues is not systematically considered. More recent studies of business participation make efforts to rectify these deficiencies. Individual participants are identified as representatives for relatively self-conscious, politically organized business groups. These individuals, not necessarily business-men themselves, are assumed to speak for segments of the local business community, whether the growth-dependent downtown interests (Molotch, 1976), the national corporate sector (Friedland, 1982; Mollenkopf, 1976), or a national capitalist class (Ratcliff, 1980; Domhoff, 1978). Both the incidence and efficacy of individual business participation may be related to the segment of the business community with which one is affiliated.

Some recent studies of business participation also question the emphasis on the resolution of public issues. Thus, the role of business participation in setting agendas—producing some local issues and preventing the production of other local issues—becomes an empirical question in its own right. Crenson (1971), for example, found that reputed influence of industry acted to suppress the emergence of air pollution as a city issue. Domhoff (1978), in his reanalysis of urban renewal politics in New Haven, documents the role of locally situated members of a national capitalist class in formulating and initiating urban renewal as both an issue and a particular kind of policy debate. These studies indicate that business participation is often instrumental in shaping a broad policy consensus in which later decisional conflicts are fought out. Thus, for example, to Domhoff, the lack of policy conflict noted by Dahl was not the result of endogenous consensus, but one that was produced by active corporate political participation.

ORGANIZATIONAL STRUCTURE AND BUSINESS POWER

Subsequent to the original concern with business political participation, there emerged a large body of comparative urban research which transformed business political domination into a politically neutral structural attribute of the urban political system. The original theoretical concern with business domination of urban policy was rendered benign as a diffusion or decentralization of power. Cities vary systematically in the number of different individuals who participate in policy issues and the extent to which the composition varies across issues. Clark (1968), for example, asked a set of informants in comparable positions (such as the mayor, president of the largest bank, and editor of the largest newspaper) to name key political participants on four issues (urban renewal, mayoral election, air pollution, War on Poverty). He argued that the larger the number of participants and the less participants overlap between issues, the more "decentralized" is the structure of influence.[2]

Thus, the political power of business slipped from view as a theoretical concern. In its place rose the structure of "community leadership" (see Grimes et al., 1976, for a recent example). The city was viewed as a political subsystem in which the patterning of political participation affects policy outputs. In general, it was argued that more decentralized leadership arrangements lead to higher levels of policy output, given the larger number of centers of power to which access may be made or from which needs may be identified (Clark, 1968; Aiken, 1970).

This new body of research treated the level of business participation as an outcome of the organizational structure of the local economy. The impact of business on local policy was seen to be contingent on the organizational structures from which business participants are drawn. Studies indicated that both the probability of participation and the influence of businessman are outcomes of the organizational role the individual occupies. An individual's reputation for community leadership was found to be related to the pattern of participation of the organization to which he/she belonged (Freeman et al., 1963). Perrucci and Pilisuk (1970) found that participation in community issues and reputation for power are greater among those who occupy executive roles in many local organizations. They argue that the locus of power is not in individual political participation, but the local structure of interorganizational relations. The political participation, of individual businessmen is not the primary attribute which accounted for their influence. Rather, in their view, the organizational structure of business is seen as the primary determinant of public policy. But in moving from individual business participation to the organizational structure of business, the conceptualization of political power changes. Here political power is seen as an attribute of the political subsystem—a capacity to identify needs, mobilize resources, and attain goals—rather than the Weberian conception of a probability that an individual or group realizes its will against opposition. In this view, the organizational structure of the local economy determines not only the pattern of business participation in local policy but also the ability of the political subsystem to adopt and implement policies of any kind.

The effects on local power structure of four organizational attributes of the local economy have received attention: differentiation, absentee ownership, extralocal linkage, and coordinative dominance. We review each in turn.

The theoretical importance of organizational differentiation of the local economy for political power derives from a democratic elitist political theory. Democratic elitist theory assumes that organization is necessary for political access and that political conflicts are primarily between organizational elites. Business organizations are assumed to compete for political influence, a competition which parallels their economic competition for factors of production and markets. From this perspective, then, the existence of a single firm which dominates the local economy must have centralizing consequences for the local polity. Conversely, a multiplicity of firms will have decentralizing consequences.[3]

In his conparative study of U.S. cities, Aiken found that organizationally differentiated local economies have more diffuse power structures and higher levels of policy outputs that ostensibly benefit poor and unorganized residents: public housing, War on Poverty, and Model Cities (Aiken 1970: 496, 508; but see Clark, 1968). Aiken and Alford (1970) argue that in organizationally differentiated economies, the large number of interorganizational linkages provides a large number of points around which political coalitions can form. Because policies can be adopted and implemented with a minimum of organizational activation, the multiplication of organizational units makes it easier to establish these "minimum coalitions." Finally, in a comparative case study of two equally polluted steel towns, Crenson (1971) found that dominance by a single corporation prevents the early emergence of air pollution as a political issue.

A second organizational attribute of the local economy which received attention was absentee ownership—local plants controlled by firms headquartered elsewhere. Research on absentee-owned plants found that their executives tend to withdraw from local political participation (Mills and Ullmer, 1946; Schulze, 1958; Noland, 1962; French, 1970). Typically, it was argued that the low level of political participation by corporate executives of nonheadquarter plants decentralizes the local structure of influence. In cities whose economies are absentee-owned, economic power is severed from political power. In his study of Ypsilanti, where Ford and General Motors' executives were found to have low levels of political involvement, Schulze (1958) argued that political and economic power had been "bifurcated" as indicated by the lack of political participation by the corporate elite. Because those with most economic power withdraw from local politics, the local political system is democratized (French, 1970). Comparative studies found that absentee ownership is associated with more diffuse local power structures (Aiken, 1970; Walton, 1967).

That resident executives of absentee-owned firms have lower levels of political participation than those of local firms is not surprising. Absentee executives have their eyes on career lines in the larger corporations. Social mobility within the firm is accompanied by geographical mobility to new firm locations. Consequently, their interest in local politics is likely to be attenuated (Phelan and Pozen, 1973: 180). The political autonomy of absentee executives is likely to be circumscribed by policy decisions of the national office which often determine local branch plant political action (Friedland, 1982; Mott, 1970). Finally, branch plant executives lack the political support provided by the large pool of corporate elites that are found in national or regional office sites.

Given the theoretical importance of absentee ownership in comparative urban power structure research and the growing recognition of the extralocal structures of which local establishments are part (Walton, 1967; Warren, 1963), it is surprising that research on the political impact of locally headquartered, national economic organizations is limited. Little research has been conducted on the political impact of local control of the economy by national firms.[4]

Finally, the relative number of control or coordinative economic *functions* has been suggested as an organizational attribute affecting a city's power structure. Hawley argued that cities whose economies have relatively few people exercising coordinative or managerial economic roles have a higher level of "systemic" power, or capacity to mobilize politically. Hawley used the ratio of city employment in managerial, official, or proprietor categories relative to the total employed labor force of the city (the "MPO ratio") as an indicator of systemic power (Hawley, 1963). He argued that the higher this ratio, the greater the number of coordinative economic roles and the less chance for a community mobilization. Hawley found that cities with high systemic power (or low MPO ratios) have higher levels of urban renewal activity. Later studies used the same MPO ratio and found that cities with low systemic power (a high MPO ratio) have centralized power structures, as measured by participational indicators (Aiken, 1970). These results are in conflict only if systemic power depends on a low level of political participation by occupants of coordinative or administrative roles. The relationship between the role structure of the local economy and patterns of political participation by the occupants of coordinative roles has not been empirically established. Later studies found that the impact of the MPO ratio was dependent on the policy studied (Smith, 1976). For areal policies—those affecting the entire jurisdiction (such as flouridation, and Hill-Burton hospital construction)—cities with high MPO ratios have higher levels of policy output. For segmental policies—those affecting particular constituencies (urban renewal, public housing—cities with high MPO ratios have lower levels of policy output.

In Hawley's terms, these results suggest that high levels of systemic power are positively associated with higher levels of policy output for segmental policy outputs, while low levels of systemic power are positively associated for areal policy outputs. In a path model where three policy variables (municipal expenditures, War on Poverty funding, and urban renewal) are used as indicators of unobserved "mobilization," Lincoln (1976) found that MPO ratio has a net negative effect on mobilization, consistent with Hawley's interpretation of low systemic power.

In summary, four organizational attributes of the local economy have been related to public policy: organizational differentiation, absentee ownership, location of national headquarters, and concentration of coordinative roles. These studies tend to neglect the differential power of the organizational units they are measuring. For example, the number of establishments is the unit of observation for constructing measures of differentiation in many studies (Aiken, 1970; Aiken and Alford, 1970; Clark, 1968). Similarly, the MPO ratio comprises all coordinative or managerial roles in the local economy. (In fact, the average social status of MPO roles is relatively low, 47 on the Duncan SEI.) This operational approach neglects the differential power of the economic organizations which own and control these establishments and hire these managers. For example, an industrial establishment owned by a small local firm may have political resources different from one owned by a national corporation whose sales are many times larger than the expenditures of the city in which it has a plant. The aggregation of organizational units must assume that these differentiated units are equally powerful in the determination of local policy.

ECONOMIC ACTIVITY AND BUSINESS POLITICAL POWER

The human ecologists defined dominance as the ability to control the conditions of existence for other units (Park, 1936). Dominant economic units are those which control the generation of employment and income growth in the local economy. In the ecological formulations, dominant economic units are those that control the exchange process with the external environment. The political implications of the ecological formulations were not clearly drawn out, but one implication is clear: dominant economic units, because they control the conditions of existence for other units, wield considerable political power. Their continued location and smooth functioning in the city are highly valued by many political participants, whether or not these are businessmen. By their control over the economic processes, they shape the material conditions necessary to the legitimacy and effectiveness of city government. The fiscal capacity and social problems of a city are shaped by the locational, production, and investment decisons of major corporations. Under such conditions, political participation by business is often as much a consequence of their political power as it is its cause. The absence of political participation by business does not necessarily indicate low levels of business political power. Control over key economic processes allows dominant businesses to shape issues and limit the parameters within which political conflicts will take place.

From this perspective the political power of business, particularly dominant economic units, derives from their locational flexibility and control of the material base of the city. Dahl (1961: 250), in his original study of New Haven, was aware of this political power. He wrote:

> Probably the most effective political action an employer can take is to threaten to depart from the community, thus removing his payroll and leaving behind a pocket of unemployed families. If the threat is interpreted seriously, political leaders are likely to make frantic attempts to make the local situation more attractive.

Various case studies have indicated the ways in which policies were reshaped to increase the profitability of the city location for certain businesses. Dahl notes two cases where urban renewal programs were tailored to prevent corporate relocation from New Haven. In another case study, Phelan and Posen (1973) documented the political power of the Greater Wilmington Development Corporation (GWDC), an organization controlled by national corporate elites, particularly those associated with DuPont. In Wilmington, an ethnically based Democratic mayor allowed this corporate group to develop blueprints for a downtown civic center, a program which displaced a large number of poor ethnic residents and closed public streets for DuPont Corporation with deleterious effects on local business (Phelan and Pozen, 1973: 244-252). Their rich case study materials indicate that urban policies are designed and redesigned to assure that the city will be a profitable location for dominant economic units.

Comparative urban research has also treated the relationship between business performance and urban policy. The studies of subnational (city and state) expenditure variation, both in the United States and Europe, find that when controls are made for variables such as industrialization or income levels, electoral variables such as turnout and party competition have few significant net effects (Dye, 1972; Fried, 1975; Hofferbert, 1971). Thus, the question was raised of whether politics matters at all in the relationship between economic activity and public policy.

The few comparative studies on urban economic structure and public expenditures seem to show that urban economic activity does affect the pattern of local public spending. Fiscal impact studies, for example, attempt to assess the demands for public expenditures generated by different types of development (Muller, 1975; Muller and Dawson, 1972). Cross-sectional studies indicate that public safety and transportation outlays are higher in cities with considerable commercial and industrial facilities, even when controlling for the higher associated property values, thus indicating increased costs or user demands (Muller, 1975; see also Kasarda, 1972). Aggregate urban economic growth generates increased local spending, and output elasticities of local expenditure are greater than one (Steiss, 1975; Heilbrun, 1974).

The relationship between economic structure and public policy has been treated as though economic development were an exogenous and nonpolitical determinant of public policy, often referred to as an "environmental" effect. The impact of economic structure on public expenditure patterns can be interpreted as evidence of the political power of business. That is, it reveals the ability of local business units to shape public expenditure and taxation in ways which are consistent with their own profitable operation. In this formulation, the political power of business is not inferred from participation in decision-making, nor from their contribution to community mobilization, but from their ability to secure a continuous stream of public goods necessary to the profitability of the urban location. Only a few studies have examined the relationship between economic structure and public policy as an indicator of the political power of business. In an exemplary study, Friedman (1977) examined water pollution control in 104 central cities and found that the more dependent the local economy was on high-polluting industries, the lower the city's participation in a federal matching grant program administered by the Environmental Protection Agency. Friedman argues (1977: 333) that such policies impose specific costs on these industries. Beyond the direct monetary costs of pollution control, the program interfaces with the control of production. Furthermore, it promotes public awareness of industrial wastes and thus increases the costs incurred in their disposal. In this case, then, the city's dependence on industries likely to be adversely affected by a particular policy impeded the development of that policy.

The ability of business to secure public goods may not depend on its continuous political participation. In fact, one might expect that business participation is inversely related to their economic power—their control over local income generation. Instead, the locational dependence of the economic unit could be an important predictor of its active political participation in local politics. This interpretation would be consistent with empirical findings that banks, newspapers, and major retail stores—which are not dominant economic units—are key participants in city politics (Molotch, 1976; Banfield and Wilson, 1963). Their economic performance depends on the investment and location decisions of other, less locationally dependent, economic units. For example, the growth of downtown retail activity has been found to be dependent on the level of downtown office activity, even when controls are made for metropolitan size, city age, and distance to nearest SMSA (Friedman, 1973; Kasarda, 1972). Increasingly, the dominant economic units are more locationally independent firms which can move between cities, if not countries. Because of their superior economic power, such dominant economic units are less dependent on political perticipation to secure necessary public policies.

Not all economic activities and units are equally dominant. A variety of models have been proposed to map the relationships between different economic sectors in a locality or region. Local economies have been differentiated in terms of basic and nonbasic activities, input-output industrial matrices, and the spatial organization of intra- and inter-organizational linkages. The economic activities of some industries and firms are more dependent on the locality and the pattern of exchanges generated there than others. The local economy is more dependent on the activities of some industries and firms than on others. So much is obvious.

But it is arguable that the differential dominance of different firms and industries in the local economy will be reflected in differential impacts on public policies. Holding all else constant, cities should be more responsive to the policy requirements of their dominant economic sectors. This differential responsiveness would indicate that different economic units/activities are more powerful than others in the determination of policy. That city policies are not equally responsive to all economic activities could be used to infer the differential power of these activities, and thus the nontechnical character of the relationship between economic structure and public policy. Little empirical work has been done of this basis. Studies indicate that the decline of downtown retail sales and housing has no effect on the level of central city urban renewal, but that the level of office activity in both the central city and the suburbs does have a positive effect (Friedland, 1982; Friedman, 1977). Again, these findings can be interpreted as indicating the relative political impact of different types of economic units/activities on urban policy.

The notion that the relationship between economic activity and public policy can be used to infer the political power of business is provocative, but incomplete. Instead of explicitly considering the impact of public policy, it simply assumes that patterns of public expenditure and taxation do indeed affect the costs, and thus the profitability, of city location. Yet what little empirical evidence there is suggests that firm location is relatively insensitive to the differential costs and benefits imposed by local taxation and spending (Brown, 1974; Morgan, 1967; Williams, 1967; Due, 1961). In general, the impact of urban spending patterns on local economic activity is not well understood (Emerson and Lamphear, 1975: 207).

Virtually all empirical research on the impact of public spending has been developed in terms of "derived demands" of the residential household. Typically, this research assumes that residential properties contain a variety of consumption items, some of which are the public goods provided at the location. These consumptions items (accessibility, public services, frontage, housing characteristics) are then regressed on property values or rents. The relationship between these consumption items and the property values or rents are argued to constitute price functions (or hedonic values), where the consumers' marginal valuations of the item equal its implicit marginal price (Deacon, 1977). While there is little evidence on the incidence of expenditure benefits (Henderson, 1977), these studies do indicate that increased public expenditures are capitalized in higher property values and rents (Yinger, 1977; Edel and Sclar, 1974; Oates, 1972). This research suggests that the public benefits are reflected in private benefits.

However, while studies have been carried out for residential consumers, there have not been comparable studies for producers. The derived demands for local public services by producers have simply not been studied (Deacon, 1977: 244-245)[5] The derivation of producer demands for public goods would indicate whether or not such goods affect different site rents within a city and thus the profitability of these locations. However, they do not speak to the issue of whether variations across cities or time in the provision and cost of public goods actually affect variations in the profitability of that location. Such research has not been carried out. The political power of business should be inferred from both the causes of public policy *and* its effects. The impact of different types of economic activity on public policy indicates the political power of business. But public policies may also affect the relative profitability of the location and thus the extent to which the interests of business are served.

BUSINESS POWER AND THE CITY: SOME COMMON PROBLEMS

We have reviewed three research traditions which have assessed the impact of various attributes of local business on local public policy, inferring the power of business from that impact. The first body of work measured the level of business participation in local decision-making processes, particularly those involving manifest political conflict. The second measured the organizational structure of the local economy and related that to aggregate levels of business political participation, on the one hand, and aggregate levels of policy outputs, on the other. The third measured the level and type of economic activity in a locality and related them to the level of policy outputs.

For different reasons, each of these research strategies has some common deficiencies. First, they all tend to neglect the actual impacts of decisions, policies, or outputs whose covariation they observe. Second, they have ambiguous criteria for the sampling of these decisions, policies, and outputs. We explain each problem in turn.

By and large, most empirical studies of the political power of business have focused on the production of local public policies rather than their consumption. The input side of policy studies has become the purview of political scientists and sociologists, while the output side has been dominated by evaluation researchers. This severance between the study of the causes of public policy and its effects has theoretical roots, but it makes it difficult for us to study some important aspects of political power.

The neglect of polic, impacts by those who studied business political participation, organizational structure, and economic activity is striking. For those who studied business participation, it was generally assumed that business political participation was necessary to reveal preferences, and that the local policy would deliver a substantive response to all groups which were successful in the use of their political resources. For those who studied the organizational structure of business, no effort was made to specify the potential congruence among the policies adopted, the benefits and costs they conferred, and the business organizations whose structures were described. This lacuna was partly because their concern was no longer with the power of business, but instead with power of the local political system to attain goals or mobilize and the related implicit assumption that public policies serve public interests. For those who studied the relationship between economic activity and public policy, the analyses have dealt with the economic causes of local policy rather than their economic consequences. In part, this results from the successful territorial claims of the economists coupled with the reluctance of the political scientists and sociologists to tread on apparently well-specified theoretical ground. But in part it derives from the assumption of the urban economists who treat the relationship between economic activity and urban public expenditure as determined by resource and technocratic restraints. Preferences for public spending are the only political variables in a world dominated by fiscal capacity and market-adjunct government functions.

This neglect of policy consequences makes it difficult to study some important dimensions of the political power of business. To what extent are policies produced which are beneficial to business without, or in spite of, their political participation? To what extent does business' political participation mark the failure to obtain a given stream of policy benefits from the local polity? To what extent are local public policies necessary to the local private economic activity of business? To what extent are continuities in public policy not the result of incremental decision-making rules of administrative and legislative elites, but instead the result of marginal adjustments made by investors and producers in the locality?

A second problem that has burdened studies of the local political power of business is that of sampling public issues or policies. The problem for all research traditions is that a theoretically cogent definition of a population of issues or policies from which to sample has not been put forward. The first obstacle has already been referred to: namely, the range of possible variation is limited by studying only those issues or policies that have managed to penetrate the public arena, precisely those issues or policies likely to be decided by public elites and to generate controversy. The role of business political power in suppressing issues or policies and/or setting agendas slips out of view. Major decisions are made all the time—the decision to relocate a plant, for example—which have major ramifications for public policy and for the life of the community as a whole. Such decisions may not become political issues. Because they are not debated by political elites, because they are not subject to an electoral mandate, and because they are accepted as the legitimate domain of the private sector do not seem scientifically valid reasons for neglecting them as the kinds of decisions which reveal, if not contain, the political power of business. The rise of no-growth coalitions in many less industrial communities in the United States, and of high levels of political conflict over the local impacts of state restructuring of industry in Europe, indicate the potential arbitrariness of deciding what is, after all, a public policy.

A second obstacle in defining a population of issues or policies is the near-universal assumption of independence between the multiple issues or policies studied. Issues and policies are treated as though they are independent and equivalent. Patterns of participation can be aggregated across them. Separate models predicting two different program expenditures can be estimated and then compared. This theoretical disaggregation of issues and policies reflects an implicit assumption that local governments tend to aggregate policy demands, and they do it in a neutral way. Thus there should be no systematic bias by nature of program. The problem is that issues and programs are not independent. Often they are functionally related. It could be argued, for example, that urban renewal programs facilitated the rise of highly skilled office economies in the central city which furthered the unemployability of the low-income resident population, while displacing many of their places of employment. In this perspective, the common determinants of urban renewal, model cities, and the War on Poverty noted by some authors is merely a reflection of program development to compensate for the social costs of the urban renewal development.

Often, issues and programs are not only interdependent but also hierarchically related. The resolution of certain kinds of issues may be exercised economically by concentration in those dominant issue or program areas. A good example can be drawn directly from Dahl's (1961) study of New Haven. Dahl documents that, despite the withdrawal of the patrician class from local political participation in the electoral arena, business elites continued to dominate the board of finance, which controlled tax structures and approval of public capital projects. These powers could certainly be used to constrain fiscal capacity and capital financing in ways which were consistent with business interests.

Finally there is the problem of the time dimension of the population of issues and policies. The tendency is for researchers to take an arbitrary time duration from which issues or programs are sampled. In many cases, these issues and programs have histories which are longer than the duration specified. In many cases, the histories of these issues and programs are quite variable. Some programs, such as urban renewal, have federal origins, date back 15 to 20 years, and are likely to become issues when site selection and capital financing arise. Other issues, such as mayoral selection, involve local institutional procedures over a century old and become issues at regular, specified intervals. The problem is that issues and programs have histories which should be analyzed in order to see from what trajectories one is sampling. Because these histories vary between policies, a researcher studying a particular set of policies or issues is aggregating across things which are functionally and politically related in the present but which are at different points in their geneses. For example, has the selection of a particular time period led to a sample of issues which involve implementation and financing decisions as opposed to initiation and formulation decisions? In other words, the stage and periodicity of an issue or program may affect the kinds of factors which determine its outcome at the juncture.

NOTES

1. Dahl, for example, used three issues: urban redevelopment, public education, and party nomination. He found that business elites were not the predominant participants, that individual businessmen had specialized patterns of participation and divergent interests according to issue area, and that public officials had the most direct influence on urban policy.

2. Bonjean (1963) and others subsequently modified this approach to identify the extent to which community leaders are visible to the community and whether they occupy legitimate leadership roles (Bonjean, 1963).

3. Aiken (1970: 495), for example, argues: "We would expect industrialized cities to have more diffused power arrangements than nonindustrialized cities. Industrialization brings about greater structural differentiation, which means that new centers of power and countervailing elites are likely to be created."

4. While not studying national economic organizations, Turk (1970, 1973a, 1973b) was first to show that a concentration of national headquarters of voluntary associations has a positive effect on the ability of a city to attract federal funds. Turk argued that empirical associations between the number of national voluntary association headquarters and per capita War on Poverty funding levels were due to the headquarters' provision of "extra-local linkages" which facilitate the flow of program information and promote national policy norms. Friedland (1982) showed that the relationship between socioeconomic conditions and adoption of federal urban policies was contingent upon the local presence of national corporate headquarters.

5. In the cost-benefit analyses of urban renewal, for example, it has been assumed that a large proportion, if not all, of development on urban renewal sites would have occurred outside the central city if urban renewal had not taken place (Weicher, 1976). Whether or not central city urban renewal did make a net contribution to the level of central city economic activity has never been empirically established.

REFERENCES

AIKEN, M. (1970) "The distribution of community power: structural bases and social consequences," in M. Aiken and P. Mott (eds.) The Structure of Community Power. New York: Random House.

——— and R. ALFORD (1970) "Comparative urban research and community decision-making." The New Atlantis 1: 85-110.

ALFORD, R. and R. FRIEDLAND (1975) "Political participation and public policy," pp. 429-479 in Annual Review of Sociology, Vol. 1. Palo Alto, CA: Annual Reviews, Inc.

APPELBAUM, R. O., J. BIGELOW, H. P. KRAMER, H. L. MOLOTCH, and P. M. RELIS (1976) The Effects of Urban Growth: A Population Impact Analysis. New York: Praeger.

APILADO, V. (1971) "Corporate government interplay: the era of industrial aid finance." Urban Affair Quarterly 7: 219-241.

BANFIELD, E. C. and J. Q. WILSON (1963) City Politics. Cambridge: Harvard University Press.

BOGUE, D. J. (1949) The Structure of the Metropolitan Community. Ann Arbor: Horace H. Rackman School of Graduate Studies, University of Michigan.

BONJEAN, C. (1963) "Community leadership: a case study and conceptual refinement." American Journal of Sociology 68: 672-681.

BROWN, D. M. (1974) Introduction to Urban Economics. New York: Academic Press.

BURT, R. (1977) "Power in a social typology," in R. Liebert and A. W. Imarshein (eds.) Power, Paradigms and Community Research. Beverly Hills, CA: Sage.

CLARK, T. N. (1968) "Community structure, decision-making, budget expenditures, and renewal in 51 american communities." American Sociological Review 33: 576-593.

CRENSON, M. A. (1971) The Un-Politics of Air Pollution. Baltimore: John Hopkins University Press.

DAHL, R. A. (1961) Who Governs Democracy and Power in an American City? New Haven: Yale University Press.

DEACON, R. T. (1978) "A demand model for the local public sector." Review of Economics and Statistics 50: 184-192.

——— (1977) "Review of the literature on the demand for public services," Papers of the National Conference on Nonmetropolitan Community Services Research, prepared for the Committee on Agriculture, Nutrition, and Forestry, U. S. Senate. Washington, DC: Government Printing Office.

DOMHOFF, G. W. (1978) Who Really Rules? New Haven and Community Power Re-examined. Santa Monica, CA: Goodyear Publishing.

——— (1967) Who Rules America? Englewood Cliffs, NJ: Prentice-Hall.

DUE, J. F. (1961) "Studies of state-local tax influences on the location of industry." National Tax Journal 14: 163-173.

DYE, T. R. (1972) Understanding Public Policy. Englewood Cliffs, NJ: Prentice-Hall.

——— (1966) Politics, Economics, and the Public: Policy Outcomes in the American States. Chicago: Rand McNally.

EDEL, M. and E. SCLAR (1974) "Taxes, spending and property values: supply adjustment in a Tiebout-Oates model." Journal of Political Economy 82: 94-154.

EMERSON, M. H. and F. C. LAMPHEAR (1975) Urban and Regional Economics: Structure and Change. Boston: Allyn & Bacon.

EPSTEIN, E. M. (1969) The Corporation in American Politics. Englewood Cliffs, NJ: Prentice-Hall.

FREEMAN, L. C. (1968) Patterns of Local Community Leadership. Indianapolis: Bobbs-Merrill.

FREEMAN, T., J. FARARO, W. BLOOMBERG, Jr., and M. M. SUNSHINE (1963) "Locating leaders in local communities: a comparison of some alternative approaches." American Sociological Review 28: 791-798.

FRENCH, R. M. (1970) "Economic change and community power structure: transition in Cornucopis," in M. Aiken and P. E. Mott (eds.) The Structure of Community Power. New York: Random House.

FRIED, R. C. (1975) "Comparative urban performance," in F. I. Greenstein, and N. W. Polsby (eds.) Handbook of Political Science, Vol. 6. Reading, MA: Addison-Wesley.

FRIEDLAND, R. (1982) Power, Crisis and the Central City. London: Macmillan.

FRIEDMAN, J. (1977) "Central business district and residential urban renewal: response to undesired change?" Social Science Quarterly 58: 45-59.

——— (1973) "Variations in the level of central business district retail activity among large U. S. Cities: 1954 and 1967." Land Economics 49: 329-35.

GALASKIEWICZ, J. (1976) Social Networks and Community Decision-Making: A Study of Corporate Actors. Ph.D. dissertation. University Chicago.

GARDINER, J. and D. J. OLSON (1974) Theft of the City. Bloomington: Indiana University Press.

GRIMES, M. D., C. M. BONJEAN, J. L. LYON, and R. L. LINEBERRY (1976) "Community structure and leadership arrangements: a multidimensional analysis." American Sociological Review 41: 706-725.

HAWLEY, A. (1971) Urban Society: An Ecological Approach. New York: Ronald.

——— (1963) "Community power and urban renewal success." American Journal of Sociology 68: 422-431.

HAYES, E. (1972) Power Structure and Urban Policy. New York: McGraw-Hill.

HEILBRUN, J. (1974) Urban Economics and Public Policy. New York: St Martin's Press.

HENDERSON, J. V. (1977) Economic Theory and the Cities. New York: Academic Press.

HOFFERBERT, R. I. (1972) "State and community policy studies: a review of comparative input-output analyses." In J. A. Robinson (ed.) Political Science Annual: An International Review 3. Indianapolis: Bobbs-Merrill.

HOLLAND, S. (1976) Capital versus the Regions. New York: St. Martin's Press.

HUNTER, F. (1953) Community Power Structure: A Study of Decision-Makers. Chapel Hill: University of North Carolina Press.

KASARDA, J. (1972) "The impact of suburban population growth on central city service functions." American Journal of Sociology 77: 1111-1124.

KESSLER, R. P. (1973) "Regional background to YBC development." University of California, Berkeley. (unpublished)

——— and C. W. HARTMEN (1973) "The illusions and the reality of urban renewal: a case study of San Francisco's Yerba Buena Center." Land Economics (November): 440-453.

LAUMAN, O., J. GALASKIEWICZ, and P. V. MARSDEN (1978) "Community structure as interorganizational linkages," pp. 455-484 in Annual Review of Sociology, Vol. 4. Palo Alto, CA: Annual Reviews Inc.

LINCOLN, J. (1977) "Organizational dominance and community structure," in R. Lie ert and A. W. Imershein (eds.) Power, Paradigms and Community Research. Beverly Hills, CA: Sage.

——— (1976) "Power and mobilization in the urban community: reconsidering the ecological approach." American Sociological Review 41: 1-15.

——— and R. FRIEDLAND (1978) "Metropolitan accessibility and socioeconomic differentiation in nonmetropolitan areas." Social Forces 57: 688-696.

MILLER, D. D. (1958) "Decision-making cliques in community power structure: a comparative study of an American and an English city," American Journal of Sociology 64: 299-310.

MILLS, C. W. and M. ULLMER (1970) "Small business and civil welfare," pp. 124-154 in M. Aiken and P. E. Mott (eds.) The Structure of Community Power. New York: Random House.

MOLLENKOPF, J. (1976) "The post-war politics of urban development." Politics and Society 28: 245-295.

MOLOTCH, H. (1976) "The city as a growth machine: towards a political economy of place." American Journal of Sociology 82: 309-331.

MORGAN, W. E. (1967) "Taxes and the location of industry." University of Colorado Studies Series in Economics, No. 4. Boulder: University of Colorado Press.

MORLOCK, L. (1964) "Business interests, countervailing groups and the balance of influence in 91 cities," in W. Hawley and F. Wirt (eds.) The Search for Community Power. Englewood Cliffs, NJ: Prentice-Hall.

MOTT, P. E. (1970) "The role of the absentee-owned corporation in the changing community," in M. Aiken and P. E. Mott (eds.) The Structure of Community Power. New York: Random House.

MULLER, T. (1975) Impacts of Land Development. Washington, DC: The Urban Institute.

——— and G. DAWSON (1972) The Fiscal Impact of Residential and Commercial Development: A Case Study. Washington, DC: Urban Institute.

MUNGER, F. (1961) "Power structure and its study," in R. Martin, et al. (eds.) Decisions in Syracuse. Bloomington: University of Indiana Press.

NADER, R. and M. J. GREEN (1973) Corporate Power in America. New York: Grossman.

NOLAND, E. W. (1962) "The roles of top business executives in urban development," pp. 226-259 in S. Chapin and S. Weiss (eds.) Urban Growth Dynamics. New York: John Wiley.

OATES, W. E. (1972) Fiscal Federalism. New York: Harcourt Brace Jovanovich.

PARK, R. (1936) "Human ecology." American Journal of Sociology 42: 1-15.

PELLEGRIN, R. J. and C. H. COATES (1956) "Absentee-owned corporations and community power structure." American Journal of Sociology 61: 413-419.

PERRUCCI, R. and M. PILISUK (1970) "Leaders and ruling elites: the interorganizational bases of community power." American Sociological Review 35: 1040-57.

PHELAN, J. and R. POZEN (1973) The Company State. New York: Grossman.

POLSBY, N. (1963) Community Power and Political Theory. New Haven: Yale University Press.

PRED, A. (1977) City-Systems in Advanced Economies. New York: John Wiley.

RATCLIFF, R. E. (1980) "Declining cities and capitalist class structure," in G. W. Domhoff (ed.) Power Structure Research. Beverly Hills, CA: Sage.

SBRAGIA, A. (1979) "Borrowing to build: private money and public welfare." International Journal of Health Services 9: 207-226.

SCHULZE, R. O. (1958) "The role of economic dominants in community power structure." American Sociological Review 23: 3-9.

SMITH, R. A. (1976) "Community power and decision-making: a replication and extension of hawley." American Sociological Review 41: 676-690.

STEISS, A. W. (1975) Local Government Finance: Capital Facilities Planning and Debt Administration. Lexington, MA: D. C. Heath.

TURK, H. (1973a) "Interorganizational activation in urban communities: deductions from the concept of system." ASA Rose Monograph. Washington, DC: American Sociological Association.

——— (1973b) "Comparative urban structure from an interorganizational perspective." Administrative Science Quarterly 18: 37-55.

——— (1970) "Interorganizational networks in urban society." American Sociological Review 35: 1-19.

WALTON, J. (1973) "The structural bases of political change in urban communities." Sociological Inquiry 43: 174-2106.

——— (1970) "A systematic survey of community power research," in M. Aiken and P. E. Mott (eds.) The Structure of Community Power. New York: Random House.

——— (1967) "The vertical axis of community organization and the structure of power." Social Science Quarterly 48: 353-368.

WARREN, R. L. (1963) The Community in America. Chicago: University of Chicago Press.

WEBER, M. (1968) Economy and Society, Vol. 1 (G. Roth and C. Wittich, eds.). New York: Bedminster Press.

WEICHER, J. C. (1976) "The fiscal profitability of urban renewal under matching grants and revenue sharing." Journal of Urban Economics 3: 193-208.

WILLIAMS, J. A. and E. ZIMMERMAN (forthcoming) "Business organization and redistributive interests in American cities." Urban Affairs Quarterly.

WILLIAMS, W. V. (1967) "A measure of the impact of state and local taxes on industrial location." Journal of Regional Science 7: 49-60.

YINGER, J. (1977) "The capitalizations of transportation costs, neighborhood amenities and local fiscal variables." Discussion Paper D-77-16. Harvard University Urban Planning, Policy Analysis and Administration.

Part III

Bureaucratic Processes and Service Delivery

7

Politicians and Urban Policy Change: The Case of Crime

STEPHEN C. BROOKS
ROBERT L. LINEBERRY

A CENTRAL ISSUE in urban political analysis—perhaps the central issue up to now—is this: What is the connection between the political process in the city and the patterns of policy change? The power structure literature, so dominant in the 1950s and 1960s, discussed at length the pattern of power in the community. But efforts to link pluralism, elitism, or some other power mix with concrete variations in policy change never seemed definitive in city case studies. Though some assessments of individual political leaders in the city—Mayor Daley and Robert Moses, for example—link their efforts to major changes in the urban landscape, such studies are often city-specific and unsystematic.

What follows is an examination of the role of urban politics and urban politicians in shaping policy change in a single issue arena: crime, a policy problem of universal importance in American cities. The research paradigm in which we work may be called the quantitative, comparative approach to urban politics and policy. It presumes that cities are best studied systematically—that is, with valid, reliable, and preferably quantitative indicators—and that larger samples produce sounder inquiries than single case studies. We add to this quantitative, comparative paradigm, however, one dimension typically missing from urban political

AUTHORS' NOTE: *This chapter is a product of the Governmental Responses to Crime Project, of which Herbert Jacob is Principal Investigator, Robert Lineberry is Co-Principal Investigator, and Anne M. Heinz is Project Manager. The research was funded by Grant 78 NI-AX-0096 from the National Institute of Justice, U.S. Department of Justice. The points of view and opinions stated here are those of the authors and do not necessarily represent the official position of the U.S. Department of Justice.*

inquiry, a longitudinal one. Almost all extant examples of the comparative, quantitative approach are cross-sectional rather than historical. Put simply, we seek to link changes in two common indicators of political change—urban elections and mayoral transitions—with changes in community crime policy. Using data from ten American cities during the period 1948-1978, we examine whether and in what ways crime rates and crime policies are sensitive to the urban electoral and mayoral cycles.

CRIME AND CITIES

Crime rates in American cities have skyrocketed since the end of World War II. In the 31 years from 1948-1978, crime rates rose from 1687 serious crimes per 100,000 people to 5109 per 100,000. During this period, the American population became thoroughly urbanized, and crime even more so. Urban politicians followed, or led, their constituents in concern about crime. Urban and even national elections turned to issues of "law and order" as politicians invested their political capital in the crime problem.

In the entrepreneurial view of urban politicians, political leaders not only respond to issues raised by their constituents but generate their own support by raising the banner for one issue or another. In this sense, politicians are "political entrepreneurs" (Frohlich et al., 1971) who capitalize on social issues to get votes (Eyestone, 1978). As crime became an important issue to Americans, it became "fair game" for the politician-entrepreneur. One veteran observer of both crime and urban politics, James Q. Wilson, remarked:

> One would suppose that politicians, in their competitive struggle for votes, would have seized upon the issue of crime almost as soon as the crime rates had started up, and, in seizing it, would have devoted their energies to outbidding each other in ways to give immediate and visible protection to its victims. Statesmen might take a longer, or at least less emotional, view of the problem; criminologists might continue their efforts to clarify its causes; but surely politicians would be quick to respond to—perhaps even exploit—rising fears of crime [1975: 64].

The urban political landscape is strewn with politicians who behaved according to this Wilsonian dictum. Frank Rizzo in Philadelphia, Tom Bradley in Los Angeles, and Charles Stenvig in Minneapolis are three police officials-turned-politician. Of the three, Rizzo and Stenvig made "law and order" their campaign linchpin.

Philadelphia and Minneapolis, in fact, are two of the ten cities upon which we draw for data on crime, politicians, and policy change. The cities we use are part of a purposive, rather than a random, sample, chosen because they exhibit regional variation, variations in their urban conditions and their political organizations and styles. The ten cities (in alphabetical order) are Atlanta, Georgia; Boston, Massachusetts; Houston, Texas; Indianapolis, Indiana; Minneapolis, Minnesota; Newark, New Jersey; Oakland, California; Philadelphia, Pennsylvania; Phoenix, Arizona; and San Jose, California. The cities vary widely in their characteristics and conditions. Bunce and Glickman (1980), for example, computed "need scores" for HUD Community Development Block Grant allocations. In a ranking of the 58 largest cities from worst-off to best-off, our cities truly run the gamut. Newark (as in most such rankings) is worst-off in socioeconomic conditions among the 58 cities; Atlanta, Boston, Oakland, and Philadelphia rank from ninth to sixteenth; Minneapolis is in the middle at 28 and Houston is near the top at forty-sixth; leaving only three—Indianapolis, Phoenix, and San Jose—which are the most socially and economically favored of the Bunce-Glickman listing.

Given, then, that the social and economic profiles of these communities vary dramatically, what linkages between political events in them and policy changes can be identified? Here we look at two events—elections and mayoral transitions—and their relationship to crime policy. First, the place of crime rates (that is perceived policy change) will be examined. Then similar questions will be asked about two policy outputs: police expenditures and police manpower levels.

PERCEIVED POLICY CHANGE

Two commonplaces of modern political science are these: (1) Politicians use issues to build coalitions and thus win elections. (2) Politician-policymakers manipulate public policies to maximize their votes at subsequent elections. A cynical interpretation of these tenets holds that political leaders who have control of policy levers will pull them to manipulate public policy or political information to their gain (Tufte, 1978).

The public gains its information about the amount of crime from published crime rates. These rates are computed by local police departments, which, for the most part, are under the jurisdiction of city politicians—mayors, city councils, and so forth. Departments must be aware of the politically sensitive nature of crime reports (Skogan, 1976).

One example comes from Washington, D.C., which law-and-order candidate Richard Nixon suggested be used as a "laboratory" to see what could be done about crime control. Within a year after Nixon's election, a new police chief, Jerry Wilson, was selected, partly because he would demand performance in reducing the crime rate. Indeed, the crime rate in Washington, D.C. did drop after Wilson became chief.

Seidman and Couzens (1974) looked closely at Washington, D.C. larcenies, an easily manipulable crime rate because officers have considerable discretion about the value of the larceny. Those over $50 go into the *Uniform Crime Reports*; those under do not. They concluded that the decrease in the crime rate was partly due to such a downgrading of larcenies. As they say, "the stated goal of reducing crime announced the desired result" (1974: 476).

Other examples show the sensitivity of crime reporting to the ebb and flow of reform. Campbell (1969) documented the sudden increase in homicide rates with O. W. Wilson's appointment as Chicago chief. In Philadelphia, a city with "notoriously unbelievable" (Skogan, 1976: 12) reporting, reform commissioner Thomas J. Gibbons in 1952 "found that for years crime records, in order to minimize the amount of crime in the city, had been faked" (Bell, 1960: 138).

We ask here, though, a rather more specific question: Do crime rates show any systematic connections with urban elections? If we presume that mayors like to show a downward trend in the rate of crime, then for them actually to lower the official crime rate, one of two things needs to happen. Either (1) the mayor would need to lean on the police department to increase arrest and apprehension rate, assuming that more aggressive policing would cut the "true" rate of crime; or (2) the police department, with or without the connivance of the mayor, would have to alter reporting/recording practices to produce less "official" crime. The former— knowing what we know about the limits of policing on crime—seems exceptionally difficult. The latter—knowing what we know about weakness of mayors and potency of bureaucracies—likewise seems no easy task.

What might be useful, then, is to create at this point a prima facie case for the supposition that elections and crime rates are linked. To do so, we turn to some evidence from two cities, Newark and Philadelphia (see Figure 7.1).

Newark. Newark is everyone's prototype of the declining eastern city. When Arthur Louis (1975) wrote about the "worst American city," Newark won hands down. As described above, Newark was ranked the neediest of 58 cities for Community Development Block Grants (Bunce and Glickman, 1980). In this troubled city, crime rates increased sixfold since the end of World War II, although not all years have exhibited similar increases. As Figure 7.1 shows, the larceny rate in Newark has actually decreased in nine of the 31 years between 1948 and 1978. Five of those nine years were years which immediately preceded a city election.

Indeed, since the end of the 1950s, every election except one has been preceded by a decline in the larceny rate.

Philadelphia. Philadelphia may be the "city of Brotherly Love," but it is also a city where "political influence, particularly mayoral influence, prevailed in the organization and operation of [the] first [police] Department, and no Philadelphia Police force since has ever been completely free from partisan party presence" (Pennsylvania Crime Commission, 1974: 71). In Philadelphia, the mayor "has traditionally exercised great influence over the Police Department, resulting in the politicization of the force and often impeding efforts to deal with corruption in the Department" (1974: 91). The culmination of police-mayoral interaction was the election in 1971 of Police Commissioner Frank Rizzo as Mayor of the City of Philadelphia.

The October 1971 Philadelphia election raised issues "not of two parties but of two Philadelphias" (Philadelphia Inquirer, October 31, 1971). A massive, 74 percent, turnout in the primary selected Rizzo as Democratic nominee, despite evidence that Rizzo personally participated in the well-known excesses of police power in Philadelphia. Relying on heavy support from the old Democratic machine, the police department, some unions, and ethnic (especially Italian) neighborhoods, Rizzo ran and won as the "law-and-order" candidate, using the slogan "Rizzo means business." The salience of crime in the 1971 election we characterize as very high, and our assessment of newspaper files ranks crime and race as the two dominant issues in the campaign.

In 1975, incumbent Rizzo ran a less heated race against a black candidate and a nominal Republican candidate. The Democratic primary was more a "Rizzo referendum" than a contest over issues, where Rizzo held almost a "messianic appeal" over his followers (Philadelphia Inquirer, May 18, 1975). Rizzo had failed a lie detector test in 1973, was accused of operating a 34-member police unit to collect information on his opponents, attacked judges who were "soft" on crime, and thundered his support for Richard Nixon. If it was a Rizzo referendum, Rizzo won handily.

Figure 1.1 indicates that larceny rates in Philadelphia were exceptionally high in the election years of 1971 and 1975; the first when Rizzo was running for mayor from the position of police commissioner, the second when he was an incumbent mayor. Between 1970 and 1971, Philadelphia larcenies increased from 12 to almost 16 per thousand population, a rate of increase unmatched before or since in Philadelphia crime reporting. The next year, Rizzo's first in office, larcenies slipped back from their election year peak to 12 per thousand. A similar pattern was repeated in 1975.

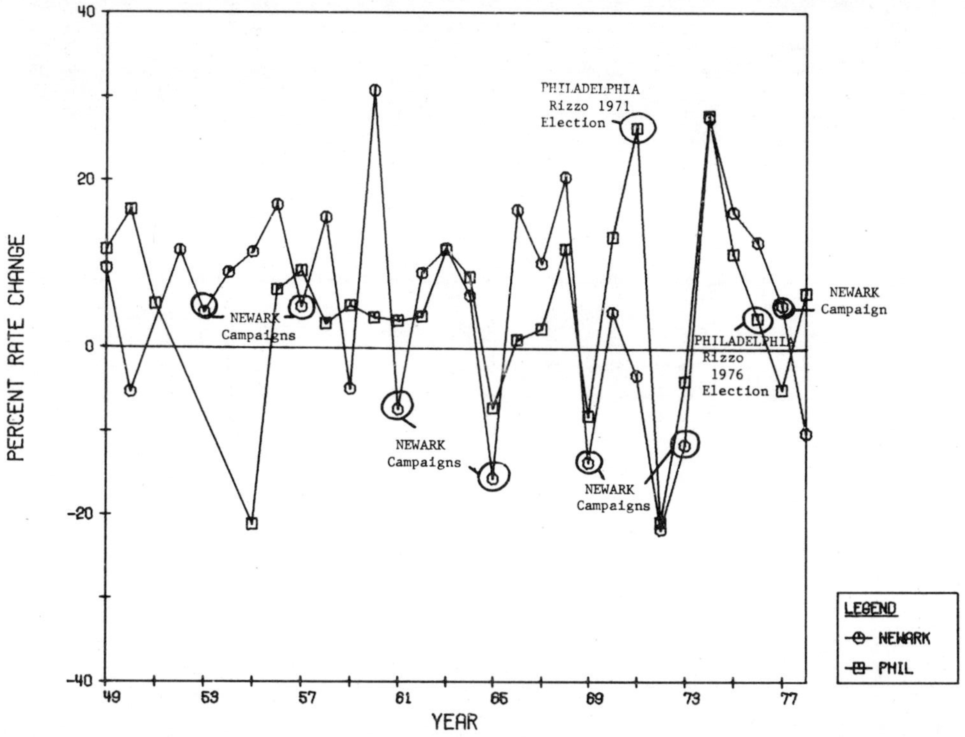

Figure 7.1 Newark and Philadelphia Larceny Rate Change

Here, then, is a curious twist. In a city where a mayoral candidate is likely to be in the strongest possible position to affect the crime rate—being himself the police commissioner—the rate of crime in an election year did not go down, but rather skyrocketed. And again in 1975, when Rizzo's influence over the police department had hardly waned, the crime rate jumped again. Perhaps no more is involved here than the long arm of coincidence. What we should do, then, is to extend our investigation beyond this tale of two cities. Possessing data on 10 cities over 31 years, we have an ample number of urban elections with which to investigate the linkages—if any exist—between crime and elections.

Hypotheses

The first set of hypotheses is based on a general hypothesis that crime rate changes will be a part of political activity in urban environments. More specifically, mayors use crime rates to help in their political position. Such a hypothesis contains two assumptions: (1) mayors have incentives and desire to use crime rates, and (2) once they have decided to use crime reports, they have the power to bring about desired results.

Making those two assumptions, conventional logic would be that as high crime rates are politically unpopular, politicians would reduce those rates during election years, leading to this hypothesis:

H1: Decreases in crime rates will occur more often in election years than in nonelection years.

This may not be always true, however. As shown in the "tale of two cities," Frank Rizzo had the incentive to increase crime rates during election years. This leads to a more general hypothesis:

H2: Dramatic shifts, whether up or down, in crime rates will occur more often in election years than in nonelection years.

Research

For this analysis *change rate* in reported crimes is the dependent variable. This avoids difficulty of enormous stability from one year to the next in crime rates. Change rate is the percentage increase in the level of a measured variable from the previous year to the measured year. Not only is this an efficient way to determine if irregular shifts took place, but it also reflects political realities. Public reporting of crime statistics most often includes a comparison of this year's crime rate compared to crime rates "one year ago." This is the same figure.

Change rates were computed for Type I FBI indexed crimes (except homicide and rape) in the ten-city sample for the 31 years 1948-1978. The mean of these change rates was used as the first dependent variable.

Larceny rates are considered the most manipulative of all crime rates. To see if other rates may have biased larceny change rates, larceny change rates are used as a second dependent variable.

Our analysis is based on ten cities over a period of 31 years providing 300 city-years as the unit of analysis. Of those 300 years, 102—or about a third—were designated election years. Because all urban elections are not held in the fall, an election year is defined as the year with a six-month recording period prior to the election. If the election was a fall election, the same year as the election is the "election year." In a city with a spring election, we used the previous year, because that is the time with the most election activity.

We tested two of these hypotheses using data on 300 city-years. Table 1.1 shows the linkage between the two dependent variables—overall crime rate changes and larceny rate changes—and presence of elections. Results are unambiguous: There is no systematic connection between the urban election cycle and changes in crime rates.

The first hypothesis suggested a relationship between decreases in crime rates and election years. Whether the mean crime rate change or larceny rate change is used, no relationship is found. Decreases in crime rates occur in 33.3 percent of the election years, as compared with 30.3 percent in nonelection years. A similar result is found for larceny rate changes, with 36.3 percent election year decreases and 32.4 percent for nonelection years. The raw chi-square and significance level of the relationship further support this conclusion.[1]

The second hypothesis is that dramatic increases or decreases would take place in election years, with more stable statistics being in nonelection years. Using an increase of 11 percent or more and a decrease of 11 percent or more as a major shift, we find similar results. For mean crime rate change, 28.3 percent of the election years and 26.8 percent of the nonelection years had major shifts. Larceny rate changes look similar, with 36.4 percent of the election years and 35.9 percent of the nonelection years showing major shifts.

Using the benefits of examining a group of cities over an extended period of time, it seems that one politically important event—an election—has no relationship to crime rates. Whatever they reflect (and that is much disputed), crime rates seem to have a life of their own. There may be many reasons why crime rates are manipulated for political purposes in specific cities. But we cannot link crime rate changes to mayoral elections in a large group of cities.

ACTUAL POLICY CHANGE

If crime rates do mirror the existence of crime and are not subject to extensive manipulation (or are actually not manipulated), crime rate

TABLE 7.1 Impact of Elections on Crime Rates

		ELECTIONS			
		Mean Crime Rate Change		Larceny Rate Change	
		Election Year	Non-election Year	Election Year	Non-election Year
NUMBER OF YEARS WITH PERCENT CHANGE GIVEN WITH COLUMN PERCENT	More than 30% decrease	0 0.0%	0 0.0%	0 0.0%	1 .5%
	21-30% decrease	0 0.0%	0 0.0%	2 2.0%	2 1.0%
	11-20% decrease	4 3.9%	5 2.5%	6 5.9%	10 5.1%
	1-10% decrease	30 29.4%	55 27.8%	29 28.4%	51 25.8%
	0-10% increase	43 42.2%	90 45.5%	36 35.3%	76 38.4%
	11-20% increase	18 17.6%	31 15.7%	21 20.6%	43 21.7%
	21-30% increase	3 2.9%	11 5.6%	6 5.9%	12 6.1%
	More than 30% increase	4 3.9%	6 3.0%	2 2.0%	3 1.5%
	Total	102	198	102	198
	Raw chi-square (Significance)	1.976 w/5 d.f. (.8525)		1.535 w/7 d.f. (.9811)	

changes may not be an appropriate dependent variable. Crime rates are outcomes of the policy process, not outputs. If a policy change comes about, it may or may not have an effect on crime rates. After all, there is little evidence that policing itself can affect the rate of crime (Kelling et al., 1974; Jacob and Rich, 1981).

Therefore, if we wish to examine crime policy, it might be better to examine policy outputs. Probably the most often used output measure is expenditures. These figures are easily found and quantified. More important, expenditures are a limited resource, and money spent on certain programs cannot be spent on others, making this measure an indicator of priorities of policymakers. Another measure of policy outputs is manpower levels devoted to tasks. In this analysis of crime policy we use both annual police expenditures and police manpower levels to detect changes in crime policy.

The most important political figure in cities is the mayor. If one conclusion can be made from a generation of research on American mayors, it is this: Mayors typically operate in a fragmented political environment and are by no means as politically powerful within their bailiwicks as are presidents or governors in theirs. Even so, some assessments of the American mayor depict mayors as relatively influential, while others relegate the mayoral role to relative insignificance. Dahl's (1961) New Haven study is in the former camp, finding a mayor-centered coalition dominating local politics. Salancik and Pfeffer (1977), for example, find that mayoral change is negligible on local budget patterns, although Welch and Karnig (1978) find that changing from white to black mayors has a modest budgetary impact.

Hypotheses

If change is going to come about in policy-making and therefore policy outputs, it would most likely be noticed in the first year of a new leadership. After making changes, the new leader becomes the incumbent and new changes become the norm (Hess, 1977; Bunce, 1980).

The history of crime control policy has shown that cities have increased their police budgets and manpower levels in an attempt to control rising crime rates. Given that, and realizing that the mayor has ultimate responsibility over police funding and manning policy, we hypothesize:

H3: Increases in police expenditures will occur more often in the first year a new mayor can have an impact than during other incumbent years.

H4: Increases in number of police officers will occur more often in the first year a new mayor can have an impact than during other incumbent years.

Research

As with crime rates, the dependent variable for this analysis was *change rate* of expenditures and number of police officers. Police expenditures were taken from the Census Bureau publication *City Government Finances* and the number of police officers came from the FBI *Uniform Crime Reports.*

The analysis uses the same 300 city-year base used earlier. Years designated "impact years" were those following election of a new mayor if the election was in the first six months of the year. For elections taking place after July 1, the impact year was designated as two years later. In the 300 city-years, we recorded 39 mayoral changes.[2] This is a surprising figure considering the stability of urban politics. In one out of every three elections, a new mayor led these cities.

Results of the analysis are shown in Table 7.2. The two dependent variables—expenditure change and number of officers change—have been categorized and shown with mayoral impact.

A new mayor seems to bring about increases in police spending.[3] Police spending increases in 94.9 percent of the impact years versus 89.6 percent of nonimpact years. Considering a one-year increase of over 11 percent substantial, results are more dramatic. In 42.1 percent of the nonimpact years such increases occurred with two-thirds (66.7 percent) of impact years having major increases. The raw chi-square of 9.25 is large but not significant. This is substantiated even more when the relationship of impact/nonimpact years is taken with the actual (noncategorized) percentage change. The Pearson correlation is .1262, with a significance level of .014.

Change in police officers provides less dramatic results but still indicates that changing mayors does have an impact on police manning policy. Number of police officers increased in 65.8 percent of the impact years versus 61.9 of the nonimpact years. Again, when we consider increases of 11 percent and more, the difference is more dramatic. Only 5 percent of the nonimpact years had major increases compared with 15.8 percent of the impact years. The raw chi-square is 16.62 and is significant with .002.

In general, we discover that changing mayors may have some impact on changing crime policy outputs. The relationships discovered are weak, but use of pooled data over time and across cities biases results toward finding no relationships. These findings support earlier research that new mayors do have impact, although not to an overwhelming degree.

TABLE 7.2 Impact of Mayoral Change on Police Expenditures and Police Force

		MAYORAL CHANGE			
		Police Expenditure Change		Number Police Officers Change	
		Impact Year	Non-impact Year	Impact Year	Non-impact Year
NUMBER OF YEARS WITH PERCENT CHANGE GIVEN WITH COLUMN PERCENT	More than 30% decrease	0 0.0%	0 0.0%	0 0.0%	0 0.0%
	21-30% decrease	0 0.0%	1 .3%	0 0.0%	0 0.0%
	11-20% decrease	0 0.0%	3 1.1%	3 7.6%	3 1.2%
	1-10% decrease	2 5.1%	23 8.8%	10 25.6%	96 36.9%
	0-10% increase	11 28.2%	124 47.5%	20 51.2%	149 56.9%
	11-20% increase	19 48.7%	81 31.0%	5 13.2%	8 3.1%
	21-30% increase	3 7.7%	17 6.5%	1 2.6%	5 1.9%
	More than 30% increase	4 10.3%	12 4.6%	0 0.0%	0 0.0%
	Total	39	261	39	261
	Raw chi-square (Significance)	9.252 w/6 d.f. (.1599)		16.621 w/4 d.f. (.0023)	

CONCLUSION

What, then, do these findings about two common political events in urban America and crime policy have to say about urban research for the next decade? While results of the analysis do not provide earth-shaking findings, we think they are suggestive of both theoretical and methodological directions for future urban research.

It has been argued that policy research is fertile ground for building urban theory because it goes beyond the argument of "who governs?" to "What difference does it make who governs?" (Masotti and Lineberry, 1976). This study has examined a most important issue for urban America—crime and what to do about it. Instead of assuming that crime policy is the result of a rational decision-making process, it investigates involvement of politics in policy-making.

Findings of this analysis which show no link between mayoral elections and crime policy but some relationship between mayoral succession and crime policy present a direction for future policy research. Much of the energy of urban research has focused on persons in power and how they got there. Political scientists, for example, examine the electoral process and structure of city governance. From these findings it would seem that a more fruitful direction of research would be to ask more about what is done by those in leadership positions instead of how they got there.

Embedded in this argument is the importance of policy change. Urban policies may vary widely from city to city, but much of that variation can logically be explained by the variation in characteristics of cities. For those interested in politics, then, the question becomes the role of politics in redirecting policy in the city.

The importance of policy change for urban research leads us to the important questions of methodology for urban research. We argued earlier that cities should be studied with a quantitative, comparative, longitudinal approach, using large samples. Research that followed showed benefits of such an approach. In the first analysis we examined the relationship of crime rates and elections by taking a comparative case study approach. It suggested that by looking at Philadelphia and Newark, crime rate changes were related to mayoral elections. We then took this interesting finding and subjected it to examination looking at 300 city-years, only to discover that such a generalization about American urban politics could not be substantiated.

The second analysis found a weak relationship between mayoral change and policy outputs. Even this weak relationship suggests a fruitful area for further research. Conducting comparative, longitudinal analysis with this size sample enables one to generalize about urban politics. A sample that treats 1950 Boston the same as 1978 Houston and finds a relationship allows one to suggest that findings can be generalized for urban politics.

This is not to suggest that urban researchers should abandon traditional case study or cross-sectional analysis. Many important foundations of urban politics come from such analysis. Research on importance of mayors in the policy process was the jumping-off point for part of our analysis. Also, the large-sample, comparative, longitudinal analysis is unable to provide the details of urban political interaction. For example, knowing that urban elections have no generalizable connection with crime policy does not discount the value of finding out why Newark elections seem to "bring about" reductions in crime rates. If we consider changes that have come about in urban America since the end of World War II, we might find it surprising to suggest that, as the city has gone through so much change, any similarities could be found between urban politics of that earlier era and now. We contend that such a hypothesis is worth testing. Ability to develop informative theory about urban politics depends on its generalizability. It is important, therefore, to examine continuity of politics of the city as it faced the challenges of earlier eras and as it faces the challenges of the 1980s.

NOTES

1. The same analysis was made on various subsamples. Election effects were examined in two time periods (1948-1962 and 1963-1978), in those cities without city managers, in incumbent elections only, and in only those cities with four-year election cycles. Although numbers changed and percentages vary, the strength of the relationship was the same as described in the text.

2. Before 1966 in San Jose, mayors were elected by the City Council from among their own. Therefore, impact years for mayoral transitions were not assigned in pre-1966 San Jose.

3. In examining policy impacts of a different type of leader—police chiefs—in these cities, we found no impact (Brooks and Lineberry, 1980).

REFERENCES

BELL, D. (1960) The End of Ideology. New York: Free Press.

BROOKS, S. C. and R. L. LINEBERRY (1980) "Mayoral transitions and crime rates: a study of ten American cities 1948-78." Presented at the 1980 meetings of the Law and Society Association, Madison, Wisconsin, June 5-8.

BUNCE, V. J. (1980) "The succession connection: political cycles and political change in the Soviet Union and Eastern Europe." American Political Science Review 74 (December): 966-976.
BUNCE, H. and N. GLICKMAN (1980) "The spatial dimensions of the community block grant program: targeting and urban impacts," chap. 2 in N. Glickman (ed.) The Urban Impacts of Federal Policies. Baltimore: Johns Hopkins University Press.
CAMPBELL, D. T. (1969) "Reforms as experiments." American Psychologist 4: 409-429.
DAHL, R. (1961) Who Governs? New Haven: Yale University Press.
EYESTONE, R. (1978) From Social Issue to Public Policy. New York: John Wiley.
FROHLICH, N., J. A. OPPENHEIMER, and O. R. YOUNG (1971) Political Leadership and Collective Goods. Princeton: Princeton University Press.
HESS, S. (1977) "Portrait of a President." Woodrow Wilson Quarterly 1 (Winter): 43-48.
JACOB, H. and M. RICH (1981) "The effects of the police on crime: a second look." Law and Society Review 15 (2).
KELLING, G., T. PATE, D. DIECKMAN, and C. E. BROWN (1974) The Kansas City Preventive Patrol Experiment. Washington, DC: Police Foundation.
LINEBERRY, R. L. (1977) Equality and Urban Policy. Beverly Hills, CA: Sage.
LOUIS, A. (1975) "The worst American city." Harpers (January): 67-71.
MASOTTI, L. H. and R. L. LINEBERRY (1976) The New Urban Politics. Cambridge, MA: Ballinger.
Pennsylvania Crime Commission (1974) Report on Police Corruption and the Quality of Law Enforcement in Philadelphia.
SALANCIK, G. R. and J. PFEFFER (1977) "Constraints on administrator discretion: the limited influence of mayors on city budgets." Urban Affairs Quarterly 12: 475-498.
SEIDMAN, D. and M. COUZENS (1974) "Getting the crime rate down: political pressure and crime reporting." Law and Society Review 8: 457-493.
SKOGAN, W. G. (1976) "Crime and crime rates," pp 105-120 in W. G. Skogan (ed.) Sample Surveys of the Victims of Crime. Cambridge, MA: Ballinger.
TUFTE, E. (1978) Political Control of the Economy. New Haven: Yale University Press.
WELCH, S. and A. KARNIG (1978) "The impact of Black elected officials on urban expenditures and intergovernmental revenue." Presented at the 1978 Annual Meeting of the American Political Science Association, New York, New York.
WILSON, J. Q. (1975) Thinking About Crime. New York: Basic Books.

8

Complex Models of Urban Service Systems

ROGER B. PARKS
ELINOR OSTROM

MOST CONVENTIONAL ANALYSES of public service delivery employ a unitary model of local governments. In such models, the "government" aggregates consumer preferences, procures and organizes means of service production, and delivers services as a monopoly supplier to constituents. Decisions about output and expenditure levels are assumed to be made by simple referenda or by omniscient and benevolent administrators. But few local government service delivery structures are so simple.

Since the early 1960s, scholars have argued for more complex models of public service delivery (for example, Ostrom et al., 1961; Margolis, 1964). Noting that the local public sector is most frequently comprised of several layers of enterprises engaging in a wide variety of exchanges, they argued the need to consider the structure of intra- and interjurisdictional arrangements as influences on service delivery. Margolis, for example, argued that the structure of interorganizational arrangements might make it possible to deal with problems that are less amenable to solution at the level of individual organizations or jurisdictions.

AUTHORS' NOTE: *We gratefully acknowledge support for the analysis presented in this chapter from the National Science Foundation in the form of Grant NSF SES 79-13397; and for the data collection from Grant NSF GI 43949. The findings and opinions expressed herein, however, are solely those of the authors and do not necessarily reflect those of the foundation. We also appreciate the helpful comments of Otto Davis, Debra Dean, Vincent Ostrom, and Barry R. Weingast on earlier drafts; and the high caliber of professional assistance we have received from the staff of the Workshop in Political Theory and Policy Analysis.*

> A consideration of the structure of governments gives a new perspective to old questions. We might ask whether some of the insoluble problems posed in the theory of public expenditures are worked out through the behavior of the structure. That is, does the structure have some of the characteristics of an industry and market, so that there is an interaction among governments which leads to desirable results? [Margolis, 1964: 236].

In addition to his concern over the neglect of interorganizational structure, Margolis also criticized analysts of public finance for excessively collapsing the internal organization of governmental units. Rather than being direct democracy or pure hierarchy, most governmental structures are far more complex. As Margolis recognized, these governmental structures may give rise to opportunities for private gain.

> Just as the market can be rigged, the government can be manipulated to protect private interests of some constituents. Just as promoters can orient and stimulate the market, there is a government bureaucracy which can gain from government activities [1964: 236-237].

Despite the cogency of Margolis's argument and those of others (such as McKean, 1964), few analysts of local service outputs and expenditures have taken into account overtly the ways the structure of intra- and interorganizational arrangements may affect the performance of local public sector economies. In order to examine the question of how internal and external decision-making structures affect performance, we draw on recent theoretical developments related to theories of public bureau behavior and how interorganizational arrangements affect public bureau behavior.

Niskanen (1971) was the first to present a rigorous and well-developed formal model of bureau behavior. Mique and Belanger (1974) criticized the Niskanen model and developed an alternative model. Orzechowski (1977) provides a cogent review of the difference between Niskanen and Mique and Belanger, and develops a third model drawing on the theoretical work of Williamson (1964) and the empirical studies by DeAlessi (1969) and Parkinson (1957). Orzechowski predicts that public bureaus, facing demands similar to private firms, will produce output at higher per-unit costs and higher labor-to-capital ratio. Orzechowski reviews several empirical studies, including his own, which compare the unit costs and the relative labor-to-capital ratios for public versus private producers. The consistent finding across a number of different studies is that unit costs of similar output are higher in public than in private firms. Public firms also employ a larger proportion of staff to capital than do private firms (see citations in Orzechowski, 1977; see also Savas, 1978; Pondy, 1968; Lee, 1972).

Work on a theory of interorganizational influences on public bureau behavior has progressed more slowly. Ostrom et al. (1961) and Ostrom and Ostrom (1965) argued for the utility of conceptualizing public service delivery structures as "industries." Public service industries, they claimed, might be analyzed using many of the same tools as those employed by economists of the industrial organization persuasion (for example, Bain, 1959). Consideration of service delivery structures in terms of their monopoly, duopoly, oligopoly, or competitive forms might enable behavioral predictions analogous to those made for private firms in market structures. In an early application of industrial organization concepts to the public sector, Bain et al. studied the water industry in northern California (1966). But little other empirical or theoretical application of industrial organization concepts occurred until the middle 1970s, partly, we believe, because of a lack of conceptual tools for characterizing the structure of service delivery arrangements in the public sector and a consequent lack of theoretically related empirical measures of this structure.

As a result of studies supported by the National Science Foundation of the organization of service delivery in metropolitan areas, two similar conceptualizations of service delivery arrangements in the public sector were developed (Ostrom and Whitaker, 1974; Ostrom et al., 1978; Savas, 1978). In both conceptualizations, service delivery arrangements are disaggregated by specific type of service (for example, general area police patrol, investigation of residential burglaries, radio communications, garbage collection, dry trash collection, and newspaper recycling). The participants in the service delivery arrangements are separately classified as *producers* of the service, as *consumers* of the service, or as *providers* or collective decision-making units that link producer and consumer.[1] Once these three types of participants are separated conceptually, they can be identified empirically for any given service in a particular geographic area (a city, a county, an SMSA, and so on). Matrices can be constructed arraying, for example, all of the producers against all of the consumers (or all groups of consumers for services with attributes of public goods). Each cell in the matrix identifies whether a service link exists between a particular producer and a particular consumer (or group) and, if so, the nature of that service link. Matrices can also be constructed for producer and provider linkages, for provider and consumer linkages, and for linkages between producers of one service and producers of other services that are necessary or useful to the former producers. These service structure matrices, together with computations based on their sizes and the patterns and types of entries, can then be used to characterize the structure of service delivery arrangements for each service of interest in many different geographic areas (see Ostrom, et al., 1978).

By analyzing the relationships between service delivery structures and the behavior of participants within structures of very different forms, we hope to improve our understanding of interorganizational influences on public bureau behavior. Does a public bureau that occupies a monopoly supply position with respect to a large population and across several different services behave differently than a set of smaller monopolists serving an equivalent total population or a mixed set of more specialized producers of particular services that, in the aggregate, supply an equivalent population? Does the availability of service supply to a given consumer (or group) from two or more different producers lead to inefficient duplication, as some would argue, or does the presence of potential competition, even if highly oligopolistic, lead to more vigorous supply efforts by all producers?

In this chapter, we develop several of the components we believe are necessary to an understanding of intra- and interorganizational influences on public service delivery. We begin by considering an output or production function for a bureau and discuss how such a function, once empirically estimated, could be used to examine bureau behavior. We then examine the technical exigencies of production functions, and consider the psychological factors likely to influence managerial decision-making in public bureaus. In a third theoretical section we discuss how different service delivery structures may affect bureaucratic supply. That is, how might the technical, psychological, and behavioral aspects of a bureaucratic supply model change from one form of service delivery structure to another? Then, in an empirical section, we develop production function estimates for some common police outputs.

BUREAU PRODUCTION FUNCTIONS

Bureaus, like firms in market situations, are involved in the conversion of inputs into one or more outputs. A number of interesting questions can be raised with respect to the conversion processes adopted by bureaus. Are bureaus technically efficient—do they obtain the maximum amount of output obtainable with a particular set of inputs? Are bureaus allocationally or price efficient—do they purchase the least cost combinations of inputs sufficient to supply a given level of output? Are bureaus scale efficient—do bureaus produce sufficient output to exhaust any scale economies in the conversion process while avoiding scale diseconomies (if any) associated with further increases in production? Finally, are bureaus efficient in a social welfare sense—do they tend to produce the level and mix of outputs preferred by the community or communities for whom their supply is intended? (These questions are raised with respect to educational bureaus by Levin, 1976. See also Leibenstein, 1976, and Moran, 1977.)

We believe that the institutional structure within and among bureaus and linking bureaus to consumer communities will have significant effects on answers to these questions. In the private sector pressures on entrepreneurs and managers resulting from the presence of many competitors are believed to lead those decision makers to choose technically and allocationally efficient production processes and input mixes. The necessity of producing outputs that consumers are willing to purchase is thought to push the levels and mixes of outputs produced toward socially efficient frontiers. We think it important to determine whether analogous pressures on decision makers can be found in public sector bureaus and, if so, how pressures for efficiency in each sense of that term used above may vary from one form of institutional structure to another. We begin this investigation by developing a bureau production function. The specific example to be used is that of bureaus supplying police services.

Police bureaus are clearly multioutput producers. (Wilson's (1968) simple trichotomy of "law enforcement, order maintenance, and service" illustratrates this, as do more elaborate typologies, such as Goldstein's (1977) eightfold list of police "objectives." In fact, three- or eightfold typologies cannot do justice to the variety of outputs that are produced by police bureaus. We will be forced to make drastic simplifications when we turn to estimating production functions for police bureaus. Initially, however, we write these functions at a very general level.

We can state an implicit police bureau production function as:

$$F(Y, X, S) = 0 \qquad [1]$$

where Y is a vector of police service outputs, X is a vector of police service inputs, and S is a vector of service condition variables. For each output, Y_i, we can write:

$$Y_i = F_i(Y_i^*, X, S) \qquad [2]$$

where Y_i^* is a vector of all outputs other than Y_i itself. This formulation explicitly acknowledges that police outputs are jointly produced, one with another. Thus, for example, arrests for automobile theft may be produced simultaneously with a reduction in the number of automobile accidents resulting from excessive speed, both by a policy of having officers stop all speeders and request a licence check on all stopped vehicles.

While the variety of police bureau outputs is very large, the variety of inputs used by most police bureaus is much less. Police bureaus are highly labor-intensive. Inputs of sworn police officers and civilian employees typically exhaust 80 to 95 percent of police agency budgets. An additional chunk of resources is devoted to the purchase (or lease) of vehicles. For purposes of specifying a production function (and, for later investigation of bureaucratic preferences), we divide sworn police officer inputs into those officers assigned to patrol duties in police agencies and those assigned to other, nonpatrol duties. With this division, we may state the production function[2] for the ith police output as:

$$0_i = K_i P^{a_i} NP^{b_i} C^{c_i} A^{d_i} \left(\prod_{j=1}^{n} S_j^{g_{ij}}\right) \left(\prod_{\substack{j=1 \\ j \neq i}}^{m} 0_j^{h_{ij}}\right) \qquad [3]$$

where

0_i = the quantity of output of the ith type produced,

K_i = a scale factor appropriate for output i,

P = the number of sworn police officers assigned to the bureau's patrol division,

NP = the number of sworn police officers assigned to nonpatrol duties in the bureau,

C = the number of civilians employed by the bureau,

A = the number of automobiles used by the department,

S_j = service conditions affecting output of the ith type, and

Y_j = all other outputs of the bureau whose production may affect the quantity of output i that is produced.

If we could assume that police bureaus operated in conditions that forced them toward efficient transformations of inputs to outputs (for example, competitive markets in local police services), we could examine questions of relative efficiency by focusing on the coefficients of this function. Holding service conditions and all other outputs constant, the coefficients a_i, b_i, c_i, and d_i for a given output would have ready interpretations. They would be the elasticities of output i with respect to their respective inputs. That is, a one percent increase in the number of sworn officers assigned to a department's patrol division, P, would be predicated to a result in a_i percent increase in the output, O_i. The marginal physical product of a particular input could be stated as the ratio of the amount of output being produced to the amount of that input employed, multiplied by the output elasticity of the input. Thus, for officers assigned to patrol, the marginal physical product could be stated as:

$$MPP = a_i \frac{0_i.}{P} \qquad [4]$$

By estimating output functions of this form for various police outputs, we could begin to examine the technical, allocational, and scale efficiency of police bureaus.

For a particular police bureau, we could enter its values for P, NP, C, and A, together with the coefficients for the estimated production function, and determine immediately whether the bureau was doing as well as predicted, or was producing more or less of a given output than predicted by the function (with other outputs held constant). Thus, we could discuss the technical efficiency of particular bureaus. If we knew the ratio of wage rates (and automobile costs) confronting a particular police bureau, we could determine whether it had chosen a least cost mix of inputs to produce a given output. The total cost function for a particular choice of inputs is:

$$TC = w_P P + w_{NP} NP + w_C C + r_A A + \sum_i r_{0_i} 0_i + FC, \qquad [5]$$

where

w_j = median wage plus wage related benefits for personnel input j,

r_A = average cost of owning and operating a police vehicle,

r_{0_i} = other variable nonwage and nonvehicle expense per unit of output type i, and

FC = fixed costs of operating a department.

If a given department were operating at the minimum total cost to produce a given mix of outputs, it should employ inputs such that the ratio of their marginal physical products (across all outputs) was just equal to the ratio of their respective wage rates. For officers assigned to patrol and officers assigned to other duties in a department, this would require that:

$$\frac{MPP_P}{MPP_{NP}} = \frac{a_i 0_i / P}{b_i 0_i / NP} = \frac{w_P}{w_{NP}}, \qquad [6]$$

or that the ratio of patrol officers to officers with other assignments in the department be:

$$\frac{P}{NP} = \frac{a_i w_{NP}}{b_i w_P}. \qquad [7]$$

Police departments that employed inputs in ratios other than these could still be operating in a technically efficient manner, but they would not have chosen the least cost combination of inputs to produce a given output.

By examining the estimated output elasticities, still assuming pressures for efficient operation, we could explore questions of scale efficiencies (holding other outputs constant). All of these elasticities are assumed to lie between zero and one. Whether the sum of the output elasticities, $a_i + b_i + c_i + d_i$, is less than, equal to, or greater than 1.0 would indicate whether production is subject to decreasing, constant, or increasing returns to scale of output. With that in mind, we could define a scale efficiency parameter, e_i, as:

$$e_i = (a_i + b_i + c_i + d_i) - 1. \qquad [8]$$

Where this parameter was positive, increasing the scale of output would be warranted on efficiency grounds. Where it is negative, decreasing the scale of operation would be in order.

Unfortunately, the factors contributing to scale efficiency are likely to be complex. Economists since at least Adam Smith have thought that economies of scale would result from the opportunities for the division of labor and corresponding specialization and mechanization. If such economies are found, then an X percent increase in all inputs should result in a greater than X percent increase in output, and the firm or bureau should expand its output accordingly. Most engineering and process estimates of production technologies indicate that returns to scale are positive over a very wide range (Walters, 1963). Thus, unless some other factor intervenes, one would expect firms and bureaus to be ever-increasing in scale.

Engineering estimates generally do not include the most commonly discussed limiting factor, however. This factor, control loss, is a managerial factor, not subject to easy engineering specification. Control loss results from the dynamics of the flow of information and control among members of the producing organization, particularly across levels of hierarchy within organizations. The information on actual operating conditions that flows upward to managers of such organizations can become distorted as lower-level personnel choose to report only those data that place their own activities in a favorable light (Tullock, 1965; Downs, 1967; Williamson, 1967). Directions and commands issued by top administrators tend to become distorted and/or treated as irrelevant to actual conditions as they flow down through hierarchical levels. Thus, control loss phenomena, which increase with an increase in the size of an organization, may serve as the limiting factor on scale economies (Coase, 1937; Robinson, 1958). The interplay, then, of technical factors contributing to economies of larger scale and of managerial factors that lead to diseconomies of larger scale will determine in specific instances whether the scale efficiency parameter, e_i, is found to be positive, negative, or zero.

Other scholars have attempted to estimate police and other public service production functions and to make efficiency comparisons along the lines indicated above (for example, Votey and Phillips, 1972; Walzer, 1972). Their findings have been mixed, some indicating the presence of scale economies (as in Walzer, 1972); others finding no such economies, or even diseconomies, in police service production. One reason for these mixed findings may be that the assumption of pressures for efficiency in production is unwarranted. In order to explore that possibility, we turn to a behavioral model of bureau supply behavior.

A MODEL OF BUREAU SUPPLY

In his formal model of bureau behavior, Niskanen (1971: 15) defines bureaus as those organizations that have both of the following characteristics:

(1) The owners and employees of these organizations do not appropriate any part of the difference between revenues and costs as personal income.

(2) Some part of the recurring revenues of the organization derives from other than the sale of output at a per-unit rate.

Most public service delivery agencies meet both these defining characteristics, in that neither managers nor employees can legally appropriate as personal income any difference (if one exists) between the lump sum revenues that are the major part of their budget and the costs of operating a department. Considerable disagreement exists about the objective function of bureau managers, but the first condition makes it unlikely that managers are motivated to reduce operating expenditures below the revenue received by a department. Most public bureaus are observed to spend as much money as they receive in appropriations. Little incentive exists to reduce expenditure levels below approved budget levels, as most rewards—status, increased salary, and other perquisites, for example—appear to be associated with larger bureau budgets. One way bureaucrats can increase the probability of larger future budgets is to ensure that all revenue allocated to them in a defined period is spent before the period is over. This line of reasoning leads quickly to assumptions of budget-maximizing and/or output-maximizing bureaus and bureaucrats. While much contemporary observation of bureaucratic behavior seems consistent with such assumptions, we prefer to delve a bit deeper into motivational concerns before stating any single bureaucratic objective.

THE BUREAUCRAT'S MAXIMAND

Unless bureau chiefs are totally under the control of providers (or consumers), at least a part of the determination of bureau output and expenditures should be responsive to their preferences. To explore the influence of bureaucratic preferences, we must consider the objective function of a bureau chief. In particular, what are the likely entries in such a function?

A brief digression on private sector enterprises is useful here. In the classical theory of firms operating in competitive markets, entrepreneur-owners are assumed to have profit maximization as their sole objective. This is plausible in light of their position as claimant to all residual profits in such a firm. This motivation is not necessary to the assumption, however. The entrepreneur in the situationally determinate environment of the competitive market (Latsis, 1972) either chooses profit maximization or market operations force his firm out of business. As Alchian (1950) demonstrated, it does not matter whether profit maximization is selected consciously or randomly; the result is the same. Only firms that follow profit maximization strategies survive. Not only is the entrepreneur forced to adopt profit maximization, but the analyst of competitive markets is also forced to assume that entrepreneurs in a competitive market are profit maximizers.

As soon as the rigorous conditions of a competitive market are relaxed, neither the entrepreneur nor the manager of a firm is forced to adopt profit maximization. Nor is the analyst forced to adopt this assumption. The theory of the management of the firm has advanced by adopting the utility maximization assumption underlying explanations of consumer behavior, rather than simple profit maximization (Marris, 1964; Williamson, 1964; Alchian, 1965). In the managerial discretion approach to private firm behavior, the manager is assumed to pursue a mixed strategy, including striving for a minimally acceptable level of profits for shareholders. Once a minimum level of profits is achieved, theorists predict the manager will trade off some of any further increments in profits for expenditures that increase the manager's status, income, and leisure. The extent to which a manager is capable of diverting resources for personal rewards "depends on the costs to the stockholders of detecting and policing the manager's behavior and effectiveness, i.e., on the costs of enforcing contracts" (Alchian, 1965: 34). These costs vary systematically with the type of ownership arrangements.

It seems reasonable to anticipate that *public sector* managers might also wish to divert some resources for personal rewards. Despite possible training and socialization with respect to "the public interest," public actors should not be thought of as a race apart from normal human beings (McKean, 1964). Utility maximization assumptions may be as useful, perhaps more so, in predicting public sector behavior as they have been in analyses of private sector managerial behavior.

Utility-maximizing private sector actors are assumed to have objective functions that include profits as well as perquisites. That is, some level of profits is necessary to the successful enterprise, with both pecuniary and nonpecuniary rewards to managers determined in part by the level of profits. Profits, too, are typically used as a measure of performance in the private sector. Firms that are more profitable than others facing comparable markets are judged more effective, and their managements are likely to receive higher rewards, both in direct salaries and in other benefits (Lewellen and Huntsman, 1970). To develop a public sector analyogy, we must define the public sector equivalent of profits.

Possible substitutes include the concept of a fiscal residuum (Orzechowski, 1977) or that of a bureau's discretionary budget (Niskanen, 1975). Orzechowski defines fiscal residuum as "the difference between tax dollars collected for a public service and the minimum costs of producing that service" (1977: 231). The bureau's discretionary budget is the difference between the maximum budget that the provider would approve for a given output and the minimum total cost of supplying that quantity of output.[3] Both concepts include some attention to the net benefits supplied to a provider's constituents. That is, we assume that no provider would approve a budget and output combination where net benefits to at least a majority of consumers are not greater than (or just equal to) zero. Otherwise, an alternative slate of candidates for provider positions could replace the current provider at the next election (see MacKay and Weaver, 1978, and Langbein, 1980, for explicit statements of this position). Using the concepts of fiscal residuum or discretionary budget, attention immediately turns to discovering whether the bureau is able to capture this surplus through overproduction or through inefficient (higher unit cost) production of a given quantity.

We prefer a different formulation, one which explicitly puts the benefits to consumers into the model. Consequently, we define a performance measure called a benefits residuum, which is the difference between the total value of a bureau's output to citizens of a providing organization and the total cost of producing that quantity of output. That is,

$$BR = \sum_i v_i 0_i - TC, \qquad [9]$$

where

BR = benefits residuum,

v_i = the average per unit valuation of output i across citizens of the providing organization, and

TC = total cost of producing the sum of the 0_i's.

Obviously, this formulation requires eventual refinement to include consideration of the distribution of benefits across consumers. We are also aware that, for most public bureaus, it is extremely difficult to obtain reliable and valid measures of outputs, of the values per unit of output, and even for the components of a cost function. The difficulty of measuring costs is not, however, avoided by using the concept of fiscal residuum.

While measuring the outputs of some public bureaus is extraordinarily difficult given the public good nature of those outputs, we feel it is essential to develop a conceptual performance measure that captures some aspects of the meaning of the term, "the public interest."

Having defined the benefits residuum as an entry in bureaucratic objective functions, we hasten to add that there are likely to be additional entries in those functions as well. Bureau chiefs are assumed capable of diverting some bureau resources to personal rewards, perhaps additional perquisites of office, increased status among his or her peers, or, like some characterizations of private sector managers, an easy life. The extent to which such capabilities are exercised depends on bureaucrats' motivations *and* the constraints they confront. We turn next to some models of bureaucratic motivation and then to constraints.

MODELS OF BUREAUCRATIC MANAGERS

Two polar models of bureaucratic managers can be advanced. One of these, a "selfless bureaucrat" model, seems implicit in much of the literature of public administration. The second polar model suggests a totally rapacious bureaucrat, who strives to capture all possible surpluses for himself and for those whom he must satisfy to maintain his position. We suspect that neither polar model captures the reality of bureaucratic motivations, but briefly review each before offering an alternative.

The selfless bureaucrat in our terminology is solely interested in maximizing the benefits residuum. The selfless bureaucrat knows citizens' preferences sufficiently well to be able to define "the public interest." He or she pursues that public interest with single-minded devotion. Using this model, one can safely dispense with controls aimed at monitoring or constraining bureau chiefs' exercises of discretion. Thus when one hears of reforms designed to strengthen hierarchical control in public bureaus or, even more, when one hears of reforms designed to reduce political control or "interference" in the operation of public bureaus, one can assume the proponents are employing a selfless bureaucrat model. By eliminating unnecessary impediments, they would argue, bureau chiefs are freed to maximize net benefits to their consumers. This selfless bureaucrat model might be viewed as the equivalent of the pure profit-maximizing model in the private sector. Given that profit maximization is not the sole strategy adopted in the private sector unless heads of firms are driven to it by the rigor of a competitive market, it seems naive to assume complete selflessness in the public sector unless comparable mechanisms exist to force such single-minded bureaucratic pursuit. We see few institutional arrangements in the public sector that carry such force.

On the other hand, total rapaciousness seems naive also. Models of this nature posit budget maximization as the sole strategy, where bureau chiefs attempt to extract the full consumer surplus available to a consuming community by increasing budgets (and output) to the maximum obtainable. We disagree with such a formulation on two different grounds. First, it seems to us to entail too much work on the part of bureaucrats and bureaucratic employees. Increasing output beyond some point, even if it were to lead to subsequently higher budgets, is likely to increase the workload of bureau managers beyond their preferred levels. Second, we perceive bureau chiefs—at least at the local level—as taking some pride in "doing a good job." Part of that good job, we believe, includes supplying net benefits to their consumers.

Because of our reservations about either polar model of bureaucratic managers, we choose to use a utility-maximizing model, with objective function entries drawn from both polar models. Our utility-maximizing model is basically a managerial discretion model, modified from the work of Williamson (1964), Pondy (1968), and Orzechowski (1977). We assume that bureau managers derive utility from the benefits residuum *and* from resources devoted to bureau personnel. The utility derived from a positive benefits residuum comes from the increased probability of job retention and advancement associated with a positive consumer surplus among the constituents of the providing organization and from personal satisfaction with serving the public well. Many local urban service bureau managers live in the community they serve and consume the output of their own bureau. We feel it is reasonable to assume that a local public service bureau chief will want to gain confidence and appreciation from citizens served and from friends, family, and neighbors for creating a positive consumer surplus. We also feel it is reasonable that a bureau chief would be willing to trade off some utility derived from the benefits residuum for increased utility derived from investment in bureau personnel.

In terms of total numbers of employees, a bureau chief's own salary and status are usually a positive monotonic function of the size of the bureau. However, the chief may derive even more satisfaction from investments in specialized personnel and staff assigned to help with the administrative load. In regard to hospitals, Lee (1972: 85) has argued that "inputs are used as status symbols, or, in other words, the pattern of input utilization defines the status group to which a hospital belongs." He also argues that hospital managers participate in a "keep up with the Jones's game" in that the "desired inputs of, say the ith hospital is assumed to be a function of the inputs utilized by other hospitals" (Lee, 1972: 85). If this same game characterizes urban police departments, which we think is the case, departments and their managers derive considerable status and recognition for investing in specialized personnel. Having their own homicide investigation bureau, bad check or arson team, dispatch facility, crime lab, and entry-level training academy adds to the status and, thus, the utility of an urban police chief. The sworn personnel assigned to administration significantly lighten the workload of a chief and also contribute to his utility.

Formally, the model we adopt posits that managers of urban police departments derive utility from both the benefits residuum and the number of staff or specialized personnel. In other words, the manager attempts to:

$$\text{Max: } U = (NP, BR) \text{ Subject to } TC<B \qquad [10]$$

The constraint is that costs cannot exceed the approved budget. Equation 10 can be rewritten:

$$\text{Max: } U = (NP, \sum_{1} v_i 0_i - w_P P - w_{np} NP - w_c C - r_a A - \sum_{i} r_{o_i} 0_i - FC) \qquad [11]$$

Figure 8.1 shows the relationships among benefits residuum, bureau output, and specialized bureau personnel. For this graphical presentation, all output variety is collapsed to a single output index. In the figure we construct iso-benefits residuum contours for all staff and output combinations. Along the ridge line (L), the marginal effects on the benefit residuum of an increase in output holding staff constant is zero. Along (D) the marginal effect on the benefits residuum of increasing staff while holding output constant is zero. At (K), where the two ridge lines intersect, the optimum staff-output combination exists in terms of the benefits residuum. By slicing through the iso-benefits residuum contour in Figure 8.1 along the ridge line (L), we obtain the possibility curve relating staff to the benefits residuum shown in Figure 8.2. Points (K)

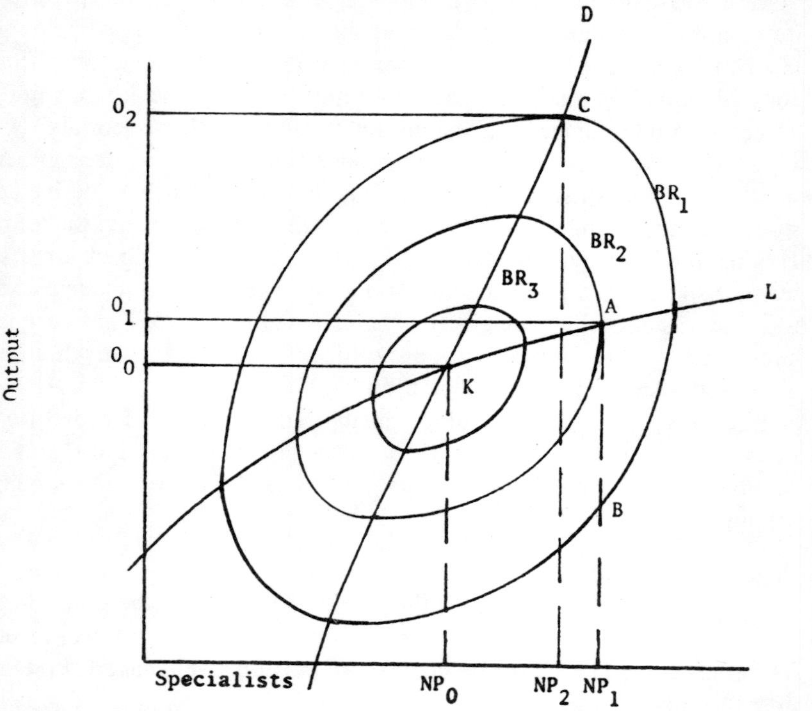

Figure 8.1 Iso-Benefits Residuum Contours

and (A) are the same in both diagrams. We can use the possibility curve and axes of Figure 8.2 to illustrate the difference between a selfless bureaucrat's objective function and preferred staff and benefits combination and the objective function and preferred position of our utility-maximizing bureaucrat.

The selfless bureaucrat would derive no utility from the employment of staff specialists per se. His or her single-minded preference for benefits residuum is illustrated in Figure 8.2 by the parallel horizontal lines labeled I_{sb1} and I_{sb2}. The line I_{sb2} represents a higher level of utility for the selfless bureaucrat and is achieved solely through increases in benefits residuum. Thus, a selfless (and efficient) bureau manager would select the output-staff combination which produces (K) on the benefits residuum. If the selfless bureaucrat were not efficient, some point below the frontier possibility curve would be chosen. In this case a reduction in the benefits residuum would result from the bureau chief's inefficiency rather than from the objectives sought by the bureau chief.

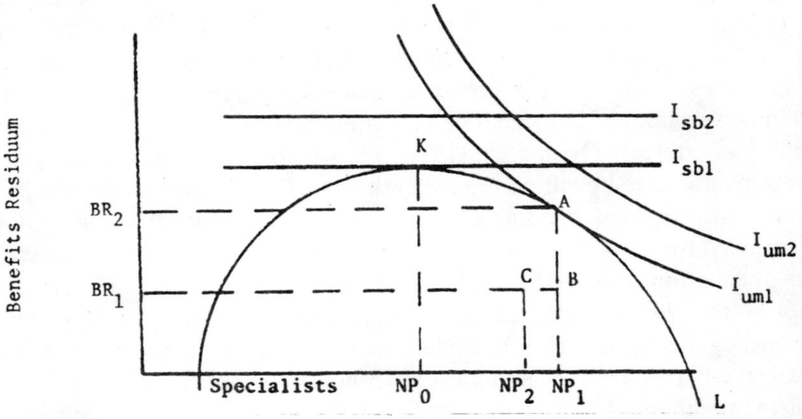

Figure 8.2 Possibility and Utility Curves

While the indifference curves of a "selfless" bureaucrat would be horizontal, the indifference curves of a utility-maximizing bureau would tip to the right and be convex to the origin, as shown by the two curves labeled I_{um1} and I_{um2}. Thus, the utility-maximizing and efficient bureau chief would select a staff-output combination at A, with a lower level of net benefits for the community. If the bureau manager were not technically efficient, more staff than NP_0 would be employed but less output would be produced. An inefficient but utility-maximizing bureau chief might end up at point (B) in both figures, where the number of staff personnel is more than NP_0 and the increase beyond optimal levels (for the community) does not even produce the most output for the combination of input resources. Any utilization of staff larger than NP_0 can be considered an investment in excess staff from the perspective of consumers. The amount of excess staff can also be considered a measure of the extent of managerial discretion that can be exercised.

The extent to which utility-maximizing bureaucrats are able to exercise their preferences without constraint is a function of the institutional arrangements in which they operate. Any point along the possibility frontier from (K) to (A), or interior points to that section of the frontier, might result from bargaining between bureaucrats and officials of providing organizations.

Bureau managers must negotiate with providers on a regular but infrequent basis (sometimes once a year) for authorization to spend a lump sum over a defined period of time. Niskanen argues that the nature of the relationship between a bureau and the officials of a providing organization frequently approximates that of a bilateral monopoly. Given that officials of providing organizations frequently have no other potential supplier of bureau services, bureau managers may gain the "same type of bargaining power as a profit-seeking monopoly that discriminates among customers or that presents the market with an all-or-nothing choice" (Niskanen, 1971: 25). If officials of the providing organization are unwilling to forego the bureau's services, they may be at a disadvantage in the negotiation over the amount of budget to be approved for a bureau. If the bureau is able to conceal information about its production and cost functions while obtaining substantial information about the demand characteristics of members of the provider's constituency, the bureau chief's capacity to confront providers with a take-it-or-leave-it proposition is enhanced (Stockfisch, 1976). This capacity is further enhanced when no competitive or potentially competitive proposals are forthcoming, either from alternative suppliers or from comparative analyses by providers of the proposals offered and accepted in other, similar situations. Of course, institutional arrangements linking the provider and the bureau (and, where applicable, other potential suppliers) will affect the relative bargaining strengths of each. The situation is not fully determinate, as it is in Niskanen's first model (1971), but rather will depend on these relative strengths (Breton and Wintrobe, 1975; Niskanen, 1975). In addition, the role of constituents or consumers as they constrain provider behavior through elections and other means must be considered in fully developed models (see MacKay and Weaver, 1978, and Langbein, 1980, for models incorporating consumers as voters.)

INTERORGANIZATIONAL ARRANGEMENTS AND BUREAU SUPPLY

Just as we think that the linkages between managers of a particular bureau, the officials of the providing organization with which bureau officials negotiate, and the citizen-consumers who are the constituents of the providing organization will have significant influences on the bureau's supply, so we believe that the interorganizational structure of service delivery in a particular area will affect the supply by all bureaus in the area. Interorganizational structure for the delivery of police services varies dramatically from area to area across the United States and also from service to service within the broad police service rubric (Ostrom et al., 1978). These diverse service delivery arrangements for policing in this country afford us fertile ground for empirically testing effects of differing interorganizational structures. In this section we first describe some of the variation in service delivery structures for policing we have found and then suggest how some structural differences might affect bureau supply.

In 1974 and 1975 we conducted a census of all organizations supplying public police services in 85 metropolitan areas (SMSAs) across the country. We found that the number of suppliers of patrol service in the 85 areas ranged from one single supplier in Meriden, Connecticut, to in excess of 90 suppliers in Paterson-Clifton-Passaic, New Jersey. Fewer than half the areas had a single supplier of patrol service for as much as 50 percent of the population. The median number of producers of general area patrol in an SMSA was 13. The existence of multiple producers of this service in the metropolitan areas meant that, at least in theory, providers could obtain comparative data to weigh against the budget and output performance of their current producing bureau and might even replace their current supplier with one or more competitors.

We posit that the number of patrol producers (multiplicity) in an SMSA should affect the level of information that citizens as as consumers and officials as providers have concerning the relative benefits residuum produced by a bureau that supplies their jurisdiction with patrol services. In a metropolitan area where there are many different producers (high multiplicity), citizens obtain information about comparative performance in several ways. Simply driving through the metropolitan area provides regular information about patrol density and the extent and style of enforcement in different jurisdictions. If a citizen or a member of his or her immediate family receives a traffic ticket in two different jurisdictions, an opportunity exists to compare directly the fairness, courtesy, and honesty of officers working in different jurisdictions. Most citizens in a metropolitan area with many jurisdictions know residents living in many different jurisdictions. Informal discussions of such personal events as being victimized, calling the police for assistance, or getting a ticket often occur among friends.

Public officials in a metropolitan area with high multiplicity also are likely to be better informed about comparative performance levels in the metropolitan area. Citizens who are unhappy with their own police and who know that their friends and neighbors receive a better level of service are more likely to call their elected officials than are citizens living in a low multiplicity area who have no way to compare the service they receive with that of other jurisdictions. Further, if city managers and/or mayors in the metropolitan area meet regularly, they can exchange relevant input and output information that helps each of them in their bargaining with police chiefs. The relative monopoly over information

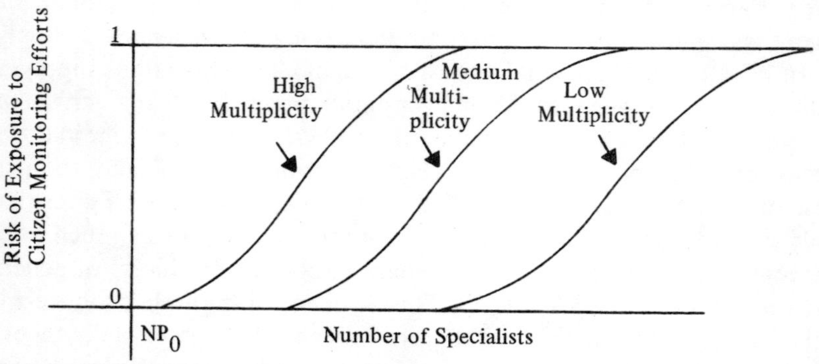

Figure 8.3 Relationship Between Risk of Exposure to Citizen Monitoring Efforts and Multiplicity of Industry Structure

that Niskanen posits is reduced in a metropolitan area with a large number of producers. Thus, police chiefs operating in metropolitan areas with high levels of multiplicity are more exposed to removal if they increase staff and other input variables beyond the level at which more effective departments in the metropolitan areas operate.

We expect that the lower costs of monitoring police department performance in metropolitan areas with high multiplicity will affect the shape of the indifference curves for bureau managers operating in those areas. In Figure 8.3 we illustrate how this might operate by examining the relationship between increasing the staff of a producing agency in any particular jurisdiction and the risk of exposure to the monitoring activities of citizens and provider organization officials. As the risk of exposure increases, bureau managers confront such sanctions as severe criticism in the local press or by the city council, or even being fired for the relatively high costs of policing in one jurisdiction when compared to similar jurisdictions in the immediate vicinity. Where the benefits residuum is increased by adding to staff, there is little or no risk of exposure to sponsor monitoring activities. However, the risk of exposure should rise exponentially once the benefits residuum maximum combination of staff and other inputs has been surpassed. We assume that this "risk" curve is affected by the level of multiplicity in the metropolitan area. In areas characterized by large numbers of other producers, the risk is higher at all levels of staff beyond the optimal number for the benefits residuum (NP_0). This contextual effect of multiplicity on the relationship between size of staff and risk of exposure to monitoring activities should be reflected in the size of staff in police departments serving similar populations in metropolitan areas varying from low to high

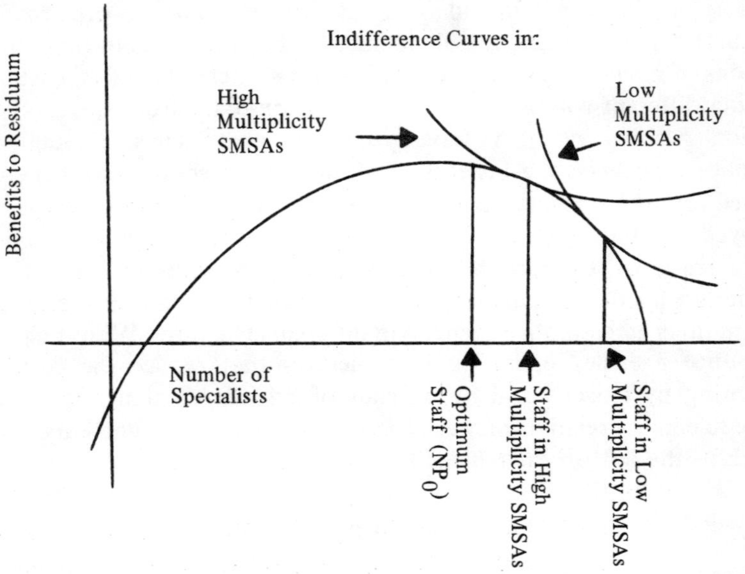

Figure 8.4 Bureau Chiefs' Indifference Curves in High and Low Multiplicity SMSAs

multiplicity. This results from the posited changes in the shape of managers' indifference curves, as shown in Figure 8.4.

Using our assumption that police chiefs are utility maximizers, Figure 8.4 shows the indifference curves for bureau chiefs in two different types of metropolitan areas (SMSAs). In low multiplicity SMSAs, the information available for monitoring a chief's performance in trading off benefits residuum and staff is low. Therefore, chiefs in such areas may be freer to indulge a preference function with a shift toward increased staff at the expense of some benefits residuum. In high multiplicity SMSAs, we posit that more information for monitoring will be available and, thus, chiefs will be forced to develop preference functions closer to that of the selfless bureaucrat, emphasizing benefits residuum more than their counterparts in areas with less information. One might argue that chiefs learn the risk of exposure and adjust their preference functions accordingly. As an alternative for those who wish to keep any individual chief's preferences constant, one can envision chiefs whose preferences are inappropriate given the level of information available for monitoring being replaced by a new top administrator whose preference function is appropriate to the situation. Police chiefs in the United States typically have quite short tenures (approximately two years on average), so this adjustment process should occur with fair rapidity.

The usefulness of utility-maximizing assumptions (as opposed to assuming a single maximand such as budget, output, or benefits residuum) depends on being able to specify the arguments in the utility functions of managers and to identify instances where managers' costs of pursuing different combinations of those entries vary from situation to situation (Alchian, 1965). We have specified two entries in such functions for police managers: the benefits residuum or net community benefits supplied by their efforts and the number of specialized personnel employed by the agency. We have also identified variations in interorganizational structure that will, we believe, provide some empirical explanation for differences in the choice of benefits residuum-specialists combinations across police agencies in different structures. We anticipate that police agencies operating in structures that reduce the cost of monitoring by citizens and by officials of provider organizations will choose to employ relatively fewer officers in specialized assignments, thus moving to the left on their benefits-specialists possibility curves.

ESTIMATING POLICE SERVICE PRODUCTION FUNCTIONS

In this section we estimate production functions for two common police outputs. We examine the functions for departments in differently structured SMSAs. We will see if the modes of production differ, as we might expect from our discussion of utility-maximizing bureau chiefs and the differences in their exposure to monitoring in different service delivery structures.

The two outputs we have chosen for analysis are (1) police response capacity, measured here by the average number of police patrol cars on the street,[4] and (2) arrests for serious crimes, measured by the number of Part I crimes cleared by arrest in a year. While police produce many varied outputs (see Goldstein, 1977), these two are certainly among the most important to consumers and to the police themselves. A high proportion of police resources are typically devoted to maintaining an on-street presence and response capacity. Commonly recommended measures of police performance usually include response time as a measure of capacity and clearances as a measure of the effectiveness of departmental investigative activities. Thus, we feel comfortable in using these output measures for our initial estimations.

ALTERNATIVE ESTIMATING TECHNIQUES

Use of the Cobb-Douglas specification for modeling the production function for response capacity may be appropriate, in that many estimated production functions have been found to resemble this specification. But using OLS estimation techniques with the logged equation and data from a large sample of departments raise other problems. It produces estimates for the coefficients of the model based on data from departments in the data set that may or may not have chosen technically efficient production strategies (efficient in the sense of maximizing output for a given set of inputs). If all departments confronted equal incentives toward production efficiency, this would not be a major problem. However, as one of our major research interests is the factors that contribute to choice of efficient production strategies, the use of average production functions is undesirable on this score.

M. J. Farrell (1957) offers a method for determining efficient production functions from observations on the inputs and outputs of many different firms. In concept the method is simple. One plots all combinations of inputs leading to a given level of output and then determines the innermost envelope of those points (the set of points closest to the origin on each of the input dimensions) for each of the isoquants of interest. Charnes, Cooper, and Rhodes (1978) offer a related method and show how it can be used with readily available linear programming algorithms.

Figure 8.5 shows a graphical representation of the method using only two inputs and one output. This restriction enables the presentation of two dimensional envelopes. Three such envelopes are shown in the figure, one for departments deploying from 1 to 5 patrol cars, a second for departments deploying 6 to 15 cars, and third for departments deploying 16 to 30 cars. These envelopes were determined by dividing the number of sworn officers assigned to patrol and the number assigned to other duties by the number of patrol cars on the street at 10 p.m., and then plotting these ratios against one another for clusters of departments with given ranges of outputs.[5] The plots were then scanned and the innermost envelopes drawn for each cluster. Within each cluster, those departments that lie to the right of the envelope employ either more sworn officers in patrol for each car on the street, more sworn officers with assignments other than patrol for each car, or a combination of more officers with both patrol and nonpatrol assignments for each car on the street.

Decreasing returns to scale are shown for response capacity output using the frontier function approach. That is, the frontier production functions represent the minimum combinations of inputs that were observed to produce given outputs. The frontiers for higher levels of output lie to the right of the frontiers for lower levels of output, thus showing the need for more than proportional increases in inputs as the scale of this particular output increases. Of course, a portion of the decreasing returns to scale in the production of response capacity shown here must be attributed to the

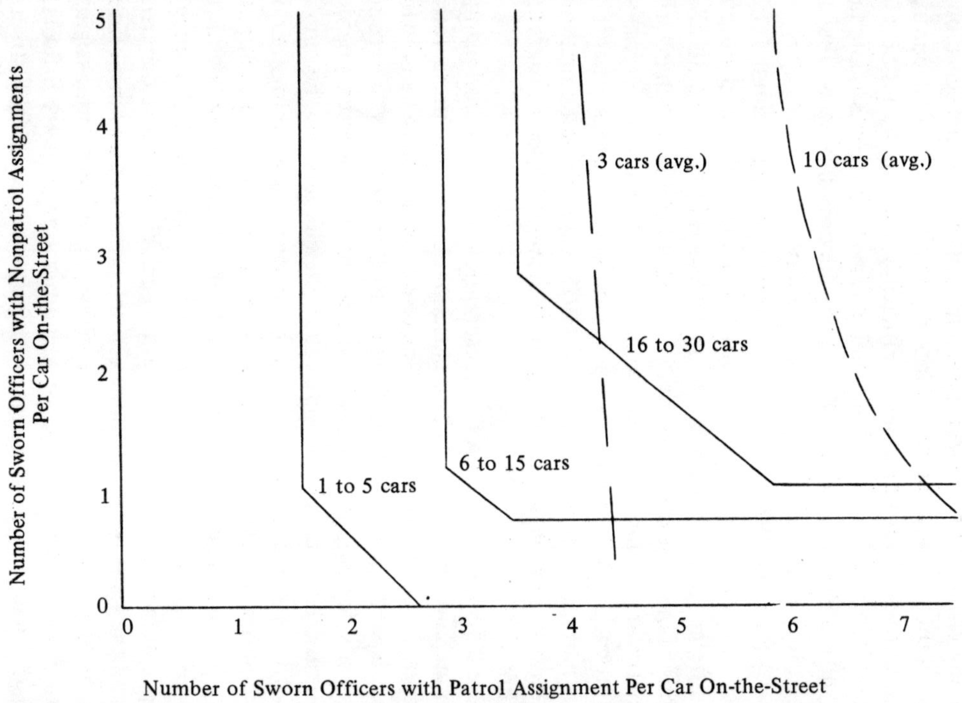

Figure 8.5 Frontier and Average Production Functions–Number on On-Street Cars

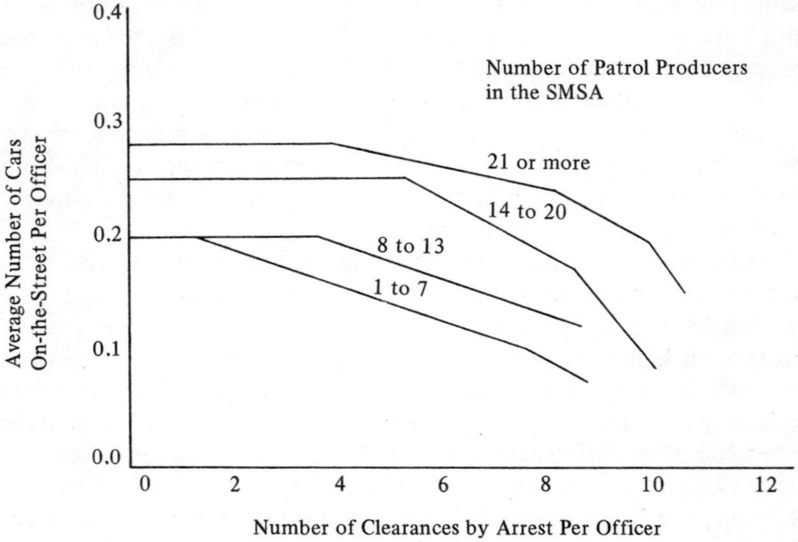

Figure 8.6 Frontier Production Possibility Curves

increased likelihood that departments are producing other types of outputs as the scale of their response capacity increases. We will investigate this phenomenon using the frontier function technique (generalized to n inputs and m outputs—see Farrell, 1957; Charnes et al., 1978) as our work in this area progresses.

Two average production function estimates are shown in Figure 8.5 for comparison purposes.[6] The average estimate for 3 cars shows that approximately 4 sworn officers with patrol assignment are required for each car on the street, a figure close to that developed in various engineering estimates of officer requirements (for example, by Kapsch, 1970, and Misner, 1960). The frontier function for 3 cars on the street indicates the possibility of achieving this output with, for example, 2.5 sworn officers per car on the street—1.5 with patrol assignment and 1 with a nonpatrol assignment. Thus, average estimates indicate a requirement of at least 12 officers to put 3 cars on the street, while frontier estimates indicate this could be done with as few as 7 to 8 officers. The average estimates for deploying 10 cars indicate a minimum of about 80 officers required, while the frontier estimates show the possibility of deploying 10 cars with approximately 40 officers. These are indeed wide differences in input requirements.

Figure 8.6 presents frontier production possibility estimates to show how metropolitan structure may, in fact, make a difference. These frontiers show the maximum combinations of clearances by arrest and cars on patrol (both standardized by the number of sworn officers) that were obtained by departments in metropolitan areas with differing amounts of multiplicity.[7] These frontiers show the trade-off possibilities for response capacity and clearances among the most efficient departments, with the normal concave shape that one expects. They also show a significant upward shift in output possibilities as the number of patrol producers in a metropolitan area increases. The most efficient producers supply more output for given inputs in high multiplicity SMSAs than do the efficient producers in lower multiplicity areas.

We believe that the frontier curves of Figures 8.5 and 8.6 are leading us to some interesting insights. It is likely that the shifts shown in Figure 8.6 result partly from the presence of more departments of smaller sizes in SMSAs with higher multiplicity. Diseconomies among the fewer (and larger) departments in low multiplicity areas could explain their curve locations. At the same time, it seems reasonable that part of the shifts is attributable to the differences in availability of information in differently structured areas. We are currently working to split these two reinforcing tendencies, attempting to specify better the effects of service delivery structures.

WHERE DO WE GO FROM HERE?

In this chapter we have proposed several models that we believe are useful in exploring police and other public agency performance. These models include an output or production function model, a model of bureau supply that suggests utility-maximizing bureau chiefs may trade off increments of net community benefits for status-raising choices of personnel deployment, and a model of some ways that interorganizational structure might influence such trade-offs. Our continuing efforts employing these models are several. At a methodological level we are developing alternative specifications for our output functions and moving toward simultaneous estimation of functions for multiple outputs. This will, we hope, give us a much better view of the scale efficiency question for agencies of differing size. In those alternative specifications we are introducing indicators for the number and different types of services supplied by the agencies so as to avoid biases introduced by their omission. We are also conducting some exploratory work comparing those departments falling close to the frontier output function estimates to those that lie closer to the average functions or to their right. This will help us to identify additional factors for improving our output model specifications, as well as providing some early evidence with respect to metropolitan structural effects.

At the theoretical level we are moving toward more complete specifications of intra- and interorganizational influences on bureau decision-making. At the intraorganizational level we will be attempting to specify likely influences from unionization, where we think this will lead to a preference for particular types of inputs and assignments, and influences from differing rank structures characterizing the relative length of chains of command. At the interorganizational level we are attempting to specify influences from additional structural factors, particularly the extent to which service supply in an area is dominated by a single bureau, and influences from the diversity of service conditions among and within jurisdictions in each area.

We have presented our models and empirical results to date as work in progress. We hope that our efforts to consider interorganizational influences will encourage other scholars to include consideration of such influences in their own theoretical and empirical work. We welcome suggestions and critiques on our work to this point as well as for further efforts.

NOTES

1. The use of producer, consumer, and provider comes from Ostrom and Whitaker (1974) and Ostrom et al. (1978) and, earlier, from Ostrom et al. (1961). Savas (1978) uses the terms provider, consumer, and arranger where we use producer, consumer, and provider.

2. We have chosen to use a Cobb-Douglas form for this initial statement of a police production function. In recent years many more sophisticated production function specifications have been developed (Intriligator, 1978: chap. 8). Most empirical estimation has, however, employed the Cobb-Douglas specification. Most readers will find such a specification more familiar. We intend to explore alternative specifications as our work continues, with particular attention to the general form of translog functions, because of their capacity to fit virtually any form (Christensen et al., 1973).

3. We use our own term, "provider," here, rather than the term "sponsor," which is more common in the literature presenting formal models of bureaucratic behavior.

4. Response capacity is actually a more complex function of the number of cars available and the volume and distribution of service requests and the time to service each. We are currently developing queueing theory based estimates of response capacity, including the number of response units and call volume estimates derived from functions representing the service conditions in each police agency's response area.

5. The method of standardization, dividing inputs by outputs, requires the assumption of constant returns to scale. We make such an assumption within somewhat arbitrary classifications of output level, but, by examining several such classifications, we are able to consider variations in scale returns.

6. These average function curves are computed using OLS estimates with equations such as 3.

7. In reality, these are "nearly maximum" combinations. In examining the plots for these relationships, some points appeared aberrant in that they were far removed from any similar points. Such aberrant points were excluded in drawing the frontiers.

REFERENCES

ALCHIAN, A. A. (1965) "The basis of some recent advances in the theory of management of the firm." Journal of Industrial Economics 14 (November): 30-41.

——— (1950) "Uncertainty, evolution and economic theory." Journal of Political Economy 58 (June): 211-221.

BAIN, J. S. (1959) Industrial Organization. New York: John Wiley.

———, R. E. CAVES, and J. MARGOLIS (1966) Northern California's Water Industry. Baltimore: Johns Hopkins University Press.

BOYDSTUN, J. E., M. E. SHERRY, and N. P. MOELTER (1977) Patrol Staffing in San Diego: One- or Two-Officer Units. Washington, DC: Police Foundation.

BRETON, A. and R. WINTROBE (1975) "The equilibrium size of a budget-maximizing bureau: a note on Niskanen's theory of bureaucracy." Journal of Political Economy 83 (1): 195-207.

CHARNES, A., W. W. COOPER, and E. RHODES (1978) "Measuring the efficiency of decision making units." European Journal of Operations Research 2 (September): 429-444.

CHRISTENSEN, L. R., W. JÖRGENSON, and L. J. LAU (1973) "Transcendental logarithmic production frontiers." Review of Economics and Statistics 55 (February): 28-45.

COASE, R. H. (1937) "The nature of the firm." Economica 4 (November): 386-405.

DeALESSI, L. (1969) "Implications of property rights for government investment choices." American Economic Review (March): 13-24.

DOWNS, A. (1967) Inside Bureaucracy. Boston: Little, Brown.

FARRELL, M. J. (1957) "The measurement of productive efficiency." Journal of the Royal Statistical Society, Series A, 120, Part 3: 253-281.

GOLDSTEIN, H. (1977) Policing a Free Society. Cambridge, MA: Ballinger.

INTRILIGATOR, M. D. (1978) "Applications to firms: production functions and cost functions," in M. D. Intriligator (ed.) Econometric Models, Techniques, and Applications. Englewood Cliffs, NJ: Prentice-Hall.

KAPSCH, S. (1970) Minnesota Police Organization and Community Resource Allocation. St. Paul: Minnesota State Planning Agency.

LANGBEIN, L. I. (1980) "Production or perquisites in the public bureaus." Presented at the 1980 meeting of the Public Choice Society, San Francisco, California, March.

LATSIS, S. J. (1972) "Situational determinism in economics." British Journal of Philosophical Science 23 (August): 207-245.

LEE, M. L. (1972) "Interdependent behavior and resource misallocation in hospital care production." Review of Social Economy 30 (March): 84-96.

LEIBENSTEIN, H. (1976) Beyond Economic Man. Cambridge, MA: Harvard University Press.

LEVIN, H. M. (1976) "Concepts of economic efficiency and educational production," pp. 149-198 in J. T. Froomkin, D. T. Jamison, and R. Radner (eds.) Education as an Industry. Cambridge, MA: Ballinger.

LEWELLYN, W. G. and B. HUNTSMAN (1970) "Managerial pay and corporate performance." American Economic Review 60 (September): 710-720.

MacKAY, R. J. and C. L. WEAVER (1978) "Monopoly bureaus and fiscal outcomes: deductive models and implications for reform," pp. 141-165 in G. Tullock and R. E. Wagner (eds.) Policy Analysis and Deductive Reasoning. Lexington, MA: D.C. Heath.

McKEAN, R. N. (1964) "Divergences between individual and total costs within government." American Economic Review 54 (May): 243-249.

MARGOLIS, J. (1964) "The structure of government and public investment." American Economic Review 54 (May): 236-242.

MARRIS, R. (1964) The Economic Theory of "Managerial" Capitalism. New York: Basic Books.

MIQUE, J. L. and G. BELANGER (1974) "Toward a general theory of managerial discretion." Public Choice 17 (Spring): 24-47.

MISNER, G. E. (1960) "Recent developments in metropolitan law enforcement." Journal of Criminal Law, Criminology and Police Science 50 (January/February): 497-508; 51 (July/August): 265-272.

MORAN, R. A. (1977) "The importance of economic criteria to agency administrators." Evaluation Quarterly 1 (February): 173-182.

NISKANEN, W. A., Jr. (1975). "Bureaucrats and politicians." Journal of Law and Economics 18 (December): 617-643.

——— (1971) Bureaucracy and Representative Government. Chicago: AVC.

ORZECHOWSKI, W. (1977) "Economic models of bureaucracy: survey, extensions, and evidence," pp. 229-259 in T. E. Borcherding (ed.) Budgets and Bureaucrats. The Sources of Government Growth. Durham, NC: Duke University Press.

OSTROM, V. and E. OSTROM (1965) "A behavioral approach to the study of intergovernmental relations." The Annals of the American Academy of Political and Social Science 359 (May): 137-146.

OSTROM, E. and G. P. WHITAKER (1974) "Community control and governmental responsiveness: the case of police in black communities," pp. 303-334 in D. Rogers and W. Hawley (eds.) Improving the Quality of Urban Management. Beverly Hills, CA: Sage.

OSTROM, E., R. B. PARKS, and G. P. WHITAKER (1978) Patterns of Metropolitan Policing. Cambridge, MA: Ballinger.

OSTROM, E., C. M. TIEBOUT, and R. WARREN (1961) "The organization of governments in metropolitan areas." American Political Science Review 55 (December): 831-842.

PARKINSON, C. N. (1957) Parkinson's Law and Other's Studies in Administration. New York: Ballantine.

PONDY, L. R. (1968) "Effects of size, complexity, and ownership on administrative intensity." Administrative Science Quarterly 14 (November): 47-60.

ROBINSON, E.A.G. (1958) The Structure of Competitive Industry. Chicago: University of Chicago Press.

SAVAS, E. S. (1978) "The institutional structure of local government services: a conceptual model." Public Administration Review 5 (September/October): 412-419.

STOCKFISCH, J. A. (1976) Analysis of Bureaucratic Behavior: The Ill-Defined Production Process. RAND P-5591. Santa Monica, CA: Rand Corporation.

TULLOCK, G. (1965) The Politics of Bureaucracy. Washington, DC: Public Affairs Press.

VOTEY, H. L. and L. PHILLIPS (1972) "Police effectiveness and the production function for law enforcement." Journal of Legal Studies 1 (June): 1-15.

WALTERS, A. A. (1963) "Production and cost functions: an econometric survey." Econometrica 31 (January-April): 1-66.

WALZER, N. (1972) "Size of operations and cost of police protection: a note." Journal of Economics and Business 54 (Spring): 60-63.

WILLIAMSON, O. E. (1967) "Hierarchical control and optimum firm size." Journal of Political Economy 75 (2): 123-138.

——— (1964) The Economics of Discretionary Behavior: Managerial Objectives in a Theory of the Firm. Englewood Cliffs, NJ: Prentice-Hall.

WILSON, J. Q. (1968) Varieties of Police Behavior. Cambridge, MA: Harvard University Press

Part IV

Citizen Preferences and Participatory Roles

9

Urban Political Cultures

DAVID KNOKE

> "A lot of the younger men wanted a nice new dark serge suit that didn't necessarily come equipped with a chauffer's cap. And the only way out was through politics; it was only when we gained a measure of political control that our people were able to come up for a little fresh air. They know that; they think of it as the big salvation for them; that's why they talk about (politics) when they all get together. It's a very serious part of the business of living."
>
> Frank Skeffington
> in *The Last Hurrah*
> (O'Connor, 1956)

The political culture concept gained wide currency more than a decade ago in the international politics and area studies field. It promised to become a master concept for ordering and making sense among a series of traditional problems such as ideology, legitimacy, sovereignty, nationhood, rule of law, and the like (Pye, 1968). The extent to which that promise was kept is not this chapter's concern. Its more modest goal is to evaluate evidence on the relevance of political culture to comparative urban political analysis in the United States. Problems remain in concept closure and measurement that have impeded the systematic application of political culture in empirical research. But we also face opportunities to move from metaphoric description to quantitative analysis with a good chance that political culture can improve our understanding of some aspects of urban political behavior.

AUTHOR'S NOTE: *I thank Terry Nichols Clark for his encouragement and suggestions on an earlier version. An NIMH research development award (K02MH00131) during this period was helpful in the preparation of this chapter.*

POLITICAL CULTURE: GENERAL DEFINITIONS

Political culture is an amorphous concept, although some consensus exists on its general substance. Almond's definition is a serviceable starting point: "attitudes toward the political system and its various parts, and attitudes toward the role of the self in the system" (Almond and Verba, 1963: 13). Pye added (1968): "a set of attitudes, beliefs and sentiments which give order and meaning to a political process and which provide underlying assumptions and rules that govern behavior in the political system. It encompasses both political ideals and operating norms of the policy." (For similar definitions that emphasize politically relevant orientations, see Putnam, 1971; Lehman, 1977; Kavanaugh, 1972; and Rosenbaum, 1975).

A political culture is distinct from a political system, although some writers include beliefs and their compatible institutional structures under the same term. A political culture is primarily a shared psychological orientation toward political objects, a manifestation of the psychological and subjective dimensions of politics. Like the general culture, a political culture is the product of a polity's history and the life history of the system's individual members. It persists through the transmission of political values and norms across generations in a socialization process (Devine, 1972: 16-17). Thus, a political culture is more enduring and fixed than are individual opinions, as noted below.

Political culture's significance lies in its consequences for political system behavior. Easton (1965: 62), for example, viewed culture as a subsystem of the external environment of the polity which "molds the constraints within which political discussion and competition take place." Using Easton's system framework, Devine (1972: 24) wrote that a political culture acts as a "screen" to filter environmental stimuli for inputs to the political system. Political cultural screens exist both at the ordinary member level, constraining the flow of political demands, and at the elite level, transforming initial demands into authoritative policies. Thus, political cultural beliefs set broad limits on acceptable demands and decisions and on the legitimate ways in which political decisions may be reached.

Most writers argue that a political culture's impact is not on specific opinions, nor is it on responses to political issues in the short run. Lehman, for example, underscored the inability of relatively durable cultural orientations to explain short-term fluctuations in political behavior. Rather, political culture specifies "the conditions under which more strategic correlations will exist in greater or lesser intensity" (Lehman, 1977: 32-33). Devine's aphorism, "culture shapes not determines," nicely summarizes this viewpoint (Devine, 1972: 35; also Elazar, 1970: 257).

CONSTITUTIVE COMPONENTS

As with many basic social science concepts, little agreement exists about the elements of political culture. Recurrent themes emphasize normative beliefs and values about various political objects (roles, institutions, incumbents, policies, the political self), rather than cognitive or affective orientations.

The following list illustrates, but does not exhaust, the set of political values which might constitute a political culture:

(1) Political Community: attachments to the political system as a whole and to other members; degree of national and local identity and belonging.

(2) The "Political" Realm: what politics is all about; boundaries between decisions which may be made by the political process and those lying outside the polity; perceptions of the political needs of society.

(3) Participation: who may participate in political activity and in what capacity; political roles, including rights and obligations; the nature of leadership and constraints on elite behavior; the proper motives of leaders and the general status of politicians; limits on mass mobilization by leaders.

(4) Power and Authority: the bases for claiming political legitimacy; limits on the use of power; modes for transforming political power resources into authority and influence, and vice versa.

(5) "Rules of the Game": proper constitutional devices for making collective decisions; the range of permissible political tactics; procedures for transfer of power from incumbents to challengers; modes for resolving conflicting interests.

(6) Political Symbols: the valued symbols of the political community, especially historical symbols; ideals and end states of the "good policy" and "good society."

(7) Scope of Government: the range of functions which government should undertake; the proper division of labor between the polity and the rest of society, and within the polity between government organizations; the amount of power desirable for government to have.

(8) Political Trust: willingness to work with various groups on collective actions; trust in representatives and in institutions to reach acceptable decisions.

(9) Issue Agenda: the range of permissible issues that should actively be considered by the political system; means for raising new issues and ordering their sequence of consideration.
(10) The Political Self: personal role in politics; sense of efficacy; political ideals and means to achieve them.

Obviously, the components of political culture can be quite complex, and many of these components may not be consciously held beliefs for many individuals. Agreement on a definitive set of constituent elements of political culture does not seem likely at present, given the different emphasis to various elements by different writers. Any efforts to measure political culture, at the individual or aggregate level, must confront decisions about which elements of culture to select for operationalization. Unfortunately, theory is not sufficiently developed at this time to offer explicit criteria for those choices.

RELATION TO PUBLIC OPINION AND IDEOLOGY

Most writers treat political culture as conceptually distinct from public opinion, although both concepts refer to attitudes toward political phenomena. Public opinion typically refers to aggregated individuals' stated preferences for a policy or course of action on a specific political issue of the day, such as busing, abortion, or presidential candidates, as well as nonpoliticized topics like child rearing or consumer goods (Monroe, 1975: 7; Weissberg, 1976:12). In contrast, political culture usually refers to more fundamental norms and values of political life, embracing deep-seated and sometimes subconscious views of the political world. Opinions on issues may be affected by factors other than political culture, such as personal interests at stake in the outcome of an issue. General cultural precepts may often conflict with specific political attitudes, as in the well-known discrepancy between support of basic civil liberties principles in abstract and unwillingness to tolerate deviant political activity (Stouffer, 1955; Prothro and Grigg, 1960; McClosky, 1964).

The distinction between political ideology and political culture is murky, a worthy subject for several books in its own right. Conceptual confusion is evident in inconsistent use of the terms "political culture," "belief systems," and "ideology." Despite these maddening inconsistencies, the common core of ideology and culture is a framework in which people order and attempt to make sense of political phenomena.

Ideologies are typically represented as highly interrelated sets of beliefs about broad-based political preferences which display internal consistency (often along a left-right or liberal-conservative dimension), high temporal stability, and frequent behavioral prescriptions (see Merelman, 1969, and Putnam, 1971, for checklists of ideological components). Little benefit would be gained in this chapter by trying to set up rigorous but arbitrary distinctions between ideology and political culture, given the prevailing confusion between the two.

URBAN SUBCULTURE TYPOLOGIES

Almond's initial use of political culture was to classify entire nations according to their dominant belief systems. The United States possesses a "civic culture" in which individual participation is highly valued and supported by existing political structures (Almond and Verba, 1963: 440). Other analysts, using less clear classification schemes, described the United States' political culture as a "liberal tradition" based on values propounded by seventeenth- and eighteenth-century political theorists (Hartz, 1955; Devine, 1172: 65). But these writers recognized internal heterogeneity even in the most fully participative nations. Persistent diversity suggests "subcultures" holding different orientations toward the inputs and outputs of the polity. Thus, the political culture concept might prove useful for research at subnational levels—in states, regions, and civil communities. Wilson (1968: 142), for example, argued that "the city is the best place to explore the kinds of attachments citizens have to the polity."

At least three typologies of American subcultures converged at a strikingly similar ideal-type classification. Williams (1961) used images of the proper role of government to classify communities into four categories (promoting economic growth; providing or securing life's amenities; maintaining traditional services; arbitrating among conflicting interests). He asserted that two distinct dimensions underlay these types: a unitary concept of the public good focusing on the substantive results, and a pluralist concept of the public good emphasizing procedural aspects of government.

Banfield and Wilson (1963; Wilson and Banfield, 1964, 1971) likewise asserted that two fundamental principles or belief systems affect issues and government structures. The first, a middle-class, Anglo-Saxon, Protestant "ethos" (unitary ethos) conceives of politics as a cooperative search for the interests of the whole community. It stresses disinterested individual involvement in civic affairs, rule by experts, and "good government." Of course, the unitary ethic may not be as disinterested and public-regarding as its partisans claim, representing instead an ideological justification for municipal reforms through which the Anglo-Saxon upper class preserved its control over cities where they had become an electoral minority (see Domhoff, 1978: 162). Banfield and Wilson's second ethic, an immigrant political culture (individualist ethos), emphasizes family needs and personal loyalties, viewing politics as a competition among individuals or groups for personal advancement.

Elazar (1970, 1972) proposed a slightly more elaborate typology of American political subcultures. Three ideal-type subcultures reflect a synthesis of two different conceptions of political order. The "commonwealth" principle emphasizes cooperation among citizens to implement certain shared moral values, a notion of common public interest. Individual interests are tempered by a general commitment to using communal power—preferably non-governmental if necessary—to intervene politically in private affairs whenever essential for collective welfare. A commonwealth principle is ascendant in a *moralistic* subculture, which closely resembles Banfield and Wilson's unitary ethos and Williams's unitary concept of the public good.

The "marketplace" principle views political activity primarily as a means for individuals and groups to better themselves economically and socially. Government serves a strictly utilitarian function of peacekeeping by balancing competing interests, with no futher obligation than preserving the marketplace itself. Lacking an overarching concept of public interest, the marketplace principle seeks to confine government action to areas, mainly economic, which encourage private initiative and widespread access to the market. This principle predominates in an *individualistic* subculture, closely resembling Banfield and Wilson's individualist ethic in character as well as in name, and Williams's pluralistic concept of the public good.

Elazar's third subculture type, the *traditionalistic*, is rooted in an ambivalence toward the marketplace and a paternal or elitist view of the public welfare. It is a survival of Southern plantation agrarianism, which accepts a hierarchical society as natural and desirable. It concedes government a political function, but one limited to maintaining existing social arrangements that confine real power to an established elite. Elazar argued that the national political culture is a complex blend of all three distinctive subcultures.

SUBCULTURE DISTRIBUTIONS

Unlike public opinion, political culture attitudes have not been top priority in national surveys. Therefore, the chief means of identifying the location of subcultures have been case studies and historical reconstructions, with a few quantitative analyses of census-type data providing additional inferences.

Elazar posited a substantial relationship between ethnoreligious groups and the three political cultures. Moralistic, traditionalistic, and individualistic subcultures originated in Old World differences among peoples coming to America over the centuries. The tendency of these groups to settle together and to move across the country in distinctive migration "streams" while keeping their political cultural beliefs relatively intact over the generations resulted in a complex "cultural geology" (Elazar, 1972: 103). Sectional (that is, regional) concentrations of specific cultural groups were overlaid by migration streams (Elazar delineated 15 major westward movements) in response to three historical frontiers (rural-land, urban-industrial, metropolitan-technological). Despite these interminglings, distinctive state political cultures persist today: The South is dominated by traditionalistic; the middle section and southwest by individualistic; and the far North, Northwest, and Pacific Coast by moralistic political cultures (1972: 119).

Within states, differential settlement patterns created local pockets of political predispositions which vary across civil communities. Elazar's *Cities of the Prairie* (1970) is painstaking, if somewhat impressionistic, effort to map the dominant political cultures of 17 medium-sized cities—many in Illinois, where the three migrant streams have a major confluence. On a broader scale, he presented classifications for 228 substate areas across the United States (1972: 106-107). Some comparative work has tried to operationalize Elazar's typology on state and substate units, usually counties, but applications at the city or individual level are lacking (see Sharkansky, 1969; Weber and Shaffer, 1972; Ritt, 1974; Elazar and Zikmund, 1975; Savage and Gallagher, 1977; Monroe, 1978).

Wilson and Banfield (1964, 1971) tried to identify the social groups most likely to adhere to the unitary and individualist ethics, using aggregated referenda voting data and a cross-sectional survey of Boston homeowners. They concluded that unitarists were more likely to be Yankees or Jews and to have higher income and education than either the individualists or those persons lacking a distinctive ethos. But less than one-fifth of the survey respondents displayed the "right" combination of attitudes on two defining subdimensions of political ethos (holistic versus localistic and community versus people-helping orientations).[1] Wilson and Banfield speculated that each ethic existed in more complete and widespread form in the past, although assimilation had not totally eradicated these differences. Further, they argued that the ethic's importance might lie in its consequences more for political elites' behavior than for ordinary citizens.

EFFECTS OF POLITICAL CULTURE

The ultimate value of the political culture concept for urban policy research depends on whether it can improve our understanding of how political systems operate. Many writers treat political culture as external to the political system. It affects political system performance by shaping political and governmental institutions and constraining the range of permissible system activities, including policy outputs. Political culture, as a system of beliefs about political objects operates both at the mass and the elite of leadership level, although the implicit assumption is that attitudes tend toward congruence among both sets of political actors.

Although political culture provides a link between individual political oreintations and the polity's behavior, the causal relationship remains obscure. While political structures may be viewed fruitfully as products reflecting political culture, structures are also important factors which may shape the political culture along certain lines. For example, a diffuse attitude syndrome may help explain why a city adopted a particular set of governmental institutions at some historical juncture. Once in place, these institutions may constrain the later electorate's attitudes to be consistent with political norms supporting these structural arrangements. The potential complexity of influence between political culture and political structure has not been adequately addressed in research (but see Lineberry and Fowler, 1967, for a recognition of the problem).

Empirical studies explicitly using a political culture framework tend to treat it as an independent variable which directly affects political behavior of entire systems, such as states, or of individuals. Elazar (1972: 129-139) demonstrated that voting turnout is highest in states with moralistic, least in states with traditionalistic, and intermediate in states with individualistic political cultures. Sharkansky (1969) found that, even controlling for region, income, and urbanization, greater state traditionalism is associated with lower political participation, less developed state bureaucracy, and less spending on a variety of social welfare programs. He argued that "culture may be most direct in its effects on popular behavior. . . . [The] impact on the forms or outputs of government is less direct, and thus more easily diluted." Monroe (1978) identified political cultural tendencies in Illinois counties using nineteenth-century national origin, religious, and political data. The cultural variables predicted 1970 referenda voting consistent with Elazar's thesis, even controlling for county demographic characteristics. Contrary to the national pattern, presence of a Southern stream within the state was associated with higher voter turnout. At least two studies of changing national political power alignments have implicitly used political culture themes to account for the rise of Republican right-wing strength in the Sunbelt states and the corresponding shift of the Yankee North to a more liberal partisan profile (Phillips, 1969; Dye, 1976: 178-186).

At the city level, Alford and Scoble (1969) analyzed structural and cultural factors affecting the policy processes of four Wisconsin cities. They concluded that public norms favoring both high bureaucratization and high participation tend to be found together (Alford and Scoble, 1969: 154). In such "modern" political systems, many demands are made and the level of decisional outputs is high. Political culture, defined as norms or value commitments of groups or the entire community apart from leadership and mass opinions, has an indirect impact on policy: "Gradually embedded in the political structures and patterns of legitimate influence in the community these norms become—in a degree which may differ sharply from community to community—accepted as the proper procedures and substance of political processes" (1969: 203). Unfortunately, this fusion of culture and structure makes their effects hard to disentangle. Like Elazar's case studies, Alford and Scoble's analysis infers the existence of a political culture from evidence on political system performance that is consistent with outcomes a given political culture might produce. As Coulter (1970) remarked on the ethos controversy, this method "tends to prove the importance of political ethos by inferentially assuming it in a tautology."

At the individual level, few studies have explicitly related political cultural beliefs to political behavior. Banfield and Wilson's (1971) Boston survey remains one of the few efforts in this vein. Holists were more likely than localists to vote for community-serving public expenditures, but "good government-minded" voters did not differ much from "benefit-minded" voters on the number or types of expenditures favored.

Political culture's impact on urban public policy may operate through elite political values as much as, if not more than, through mass attitudes. Political elites control the resources and authority ultimately required to make binding collective decisions for the community. In Laumann and Pappi's (1976) elite subsystem model, value orientations play a crucial part in determining the structure of the community affairs discussion network, which in turn affects such political outcomes as issue consensus, cleavage structure, and specific policy decisions. Another major study of political elites with a tangential relation to political culture was the massive analysis of 82 city councils in the San Francisco Bay Area (Eulau and Eyestone, 1968; Eulau and Prewitt, 1973). Their concept of city council's collective "policy map" is evocative of an elite political culture, since the map concerns "the predilections, preferences, orientations, and expectations of policy makers." More precisely, a policy map has three components, which are collective attitudes of a general nature about (1) perceived environmental problems facing the community; (2) values and goals or "images" held about the desired future of the city; and (3) recommendations for action or "policy positions"—that is, preferred outcomes emerging during the initial stages of policy determinations. Empirical findings were that each policy map component covaried in expected ways with the communities' level of policy development, suggesting that elite subculture values affect a political system's policy outcomes.

Finally, we should note the extensive research on political alienation which may be relevant to political culture (see Wright, 1976, for an overview). Feelings of personal efficacy, system efficacy, and trust in responsiveness of leaders and institutions are important political cultural beliefs which may affect political system behavior. While a comprehensive review is not possible, the main effect of efficacy is fairly well established: Low levels of personal efficacy and feelings of system nonresponsiveness covary strongly with withdrawal from political participation. The consequences of political distrust on participation are not so obvious, being contingent in complex fashion on personal efficacy and issue preferences or ideology (Miller, 1974; Aigner, 1977). As dimensions of political culture which vary across social groups and presumably between communities, efficacy and trust are important concepts for which some secondary evidence from national surveys is available.

SOME HYPOTHESES

To date, few empirical studies have directly measured American political cultural beliefs at the individual level or tried to relate variation in cultural norms to city-level political system characteristics. Thus, the research potential is enormous, although the payoff in significant findings remains unclear. For political culture to become a key explanatory concept in comparative urban research, we urgently need more precise expectations than a global assumption that political culture "shapes" political system behavior. As a first step in this direction, I collated known, suggested, and inferred relationships between urban political culture and some basic political variables, using a contrast between two fundamental types of political subcultures implicit in the dominant typologies.

These two subcultures, intended as polar types, are labeled "pluralistic" and "unitary." Like other theorists, I assume that the major difference in the political beliefs present in these cultures concerns the ultimate interests and purposes to be served by a political subsystem. A pluralistic subculture sees politics as serving private interests. The polity is an arena for contest among groups with divergent, if not opposing, interests and government is a mechanism for furthering personal and group gains. A unitary subculture sees politics in a contrasting light, as a process for maximizing collective benefits by pursuing common values concerning the good society.

The hypothesized relationships between formal institutions and political behaviors in these two subcultures are contrasted in Table 9.1. Some expectations are drawn from other writings, and some are logical extensions of arguments made by political culture theorists. Other hypotheses are more tentative, based on presumptions about covariation of political system properties. Some hypotheses may appear contradictory, but such ambiguity is inherent in a field where theoretical speculation is rife but empirial verification lags far behind. Space constraints prevent more than a superficial discussion of all the hypotheses. As an alternative, I expound at some length on the relationship between political subculture and the performance of the voluntary association, or interest group, system.

Verba and Nie (1972) identified communal contacting of government officials as a major nonelectoral mode of political participation. *Communal contacts* involve efforts to influence government decisions on issues affecting large segments of the population, in contrast to influence on personal or immediate family problems (that is, personalized contacting). In their analyses of voluntary association affiliations, Verba and Nie found that respondents' communal contact rates were strongly increased through active involvement in one or more voluntary organizations (1972: 200). Even holding constant respondents' socio-

TABLE 9.1 Hypothesized Differences in Political System Characteristics of Two Idealized Political Subcultures

Political System Characteristics	Political Subcultures: Pluralistic	Political Subcultures: Unitary
FORMAL GOVERNMENT STRUCTURE	"Unreformed" (partisan elections, wards, mayor-council)	"Reformed" (nonpartisan, at-large, manager-council)
PARTY SYSTEM	Professional politicians Cohesive parties organized on "business" or "machine" model Candidate selection by slating Politicians motivated to seek rewards of office One-party dominance	Amateur politicians Frequent personnel turnover Internal party factions Social movement aspect: ad hoc and third parties Primary system of candidate selection Office-seekers motivated by reward of policy control Competitive parties Frequent incumbent changes
ELECTION SYSTEM	High turnout rates due to disciplined party effort Durable party cleavages along population lines Few referenda on issues	Lower turnout rates due to weak parties Great variability in turnout due to response to issues Shifting political alignments within mass electorate Greater referenda use
INFORMAL ELITE DECISION-MAKING SYSTEM	Shifting coalitions across specific issues Instability of elite membership over time Noneconomic elites more influential in many decision areas Decentralized decision-making structure Opposition groups more readily mobilized Highly segmented internal differentiation of elites	Cohesive and unified elite, with few internal factions Economic elites dominant across many issues Centralized decision-making structure, especially when elite interests are involved

TABLE 9.1 Continued

Political System Characteristics	Political Subcultures	
	Pluralistic	Unitary
BUREAUCRACY	Spoils system, patronage Nonprofessional department heads Bureaucracy less responsive to citizen demands "Corruption" and inefficiency in service delivery	Merit system, civil service Professional administrators Greater production efficiency Higher responsiveness to citizen complaints (less "red tape")
VOLUNTARY ASSOCIATION SYSTEM	Organized around narrowly defined group interests Association leaders play main role in contacts with political system Groups seek benefits for their members Neighborhood groups well organized and effective Formal city recognition of neighborhood groups Decentralization (little city halls)	Many ad hoc groups organized around issues and "civic interest" Enhanced role of groups in local politics High levels of membership participation and mobilization Prevalence of citywide and uncontested associations Bias toward high-SES membership
POLICY DECISIONS	Concentration on pragmatic issues (e.g., service delivery) City executives more insulated from group demands Tendency for leaders not to initiate policy without strong outside demands Policy outputs responsive to minority group needs (non-middle-class: labor, ethnic-racial minorities) Greater reliance on supra-local government aid (state and federal) Greater innovation on controversial issues	Symbolic politics importance (e.g., moral crusades) City executives more vulnerable to group demands Leaders may initiate policy without strong outside pressure Policy outputs more responsive to middle-class and business needs Greater reliance on local government efforts Greater stalemate on attempted controversial innovations

economic status, communal contacts were greatest in rural and smaller communities, declining sharply with increased urban size, particularly in large suburbs (p. 240). These findings not only indicate the importance of voluntary associations for mobilizing nonelectoral political activity; they also imply that variation across communities may be linked to political subcultural differences.

Several components of political culture are implicated in voluntary association behavior. Norms about who should participate in political affairs, the legitimate means by which citizens should try to affect government policies, and the appropriate scope of policies to be influenced may differ between pluralistic and unitary political subcultures.

A unitary culture emphasizes collective efforts to achieve solutions to common problems facing the community, not necessarily through government action where private group intiative will suffice. Unitarists ascribe high legitimacy to voluntary associations as corporate actors, in their capacity to aggregate and articulate citizen views to officials. The political elites in a unitary subculture welcome, if not positively encourage, organized expressions of interest, both as a way to obtain information about citizen needs and as a guide to setting public policy.

In contrast, a pluralistic political culture deemphasizes collective solutions to problem-solving. Voluntary associations and interest groups are not legitimated as vehicles for pressing claims in the polity. Personalized contacting of elected and party officials about individual or family problems is the preferred mode of redress through the political system. Elite cultural norms view group efforts to influence public polity as unwarranted interference with officials, who should be left alone to make up their minds without outside pressure. Given these different cultural beliefs about the role of organized interests, different group activity patterns are anticipated in unitary and pluralistic subcultures.

In communities where unitary norms predominate, numerous issue-oriented groups should proliferate. Groups such as "taxpayers' associations" and "civic betterment" organizations will claim to speak for widely shared concerns of the whole community. These cities will contain one or more enduring citywide associations capable of building consensus and solidarity within the organizational network: "They can initiate and pursue noncontroversial projects and play mediating roles in disputes by introducing *un*disputed value inputs for partisans to employ as legitimating forces" (Turk, 1977: 145). Voluntary group members in unitary communities are especially likely to be recruited from the higher strata, not only because socioeconomic status is strongly correlated with membership, but also because unitary subcultures are more prevalent in middle-class communities. Thus, voluntary groups will more readily mobilize their middle-class bases for various political influence activities (public meeting attendance, petition drives, letter-writing, and the like). The frequency of ties between voluntary organizations will be greater, since middle-class persons have more multiple-organization affiliations (McPherson, 1978), and also because unitary cultural norms encourage interorganizational coalitions for advancement of collective welfare.

Communities dominated by pluralistic subcultural values probably contain fewer organized interest groups. The existing groups differ from the types in unitary cultures. Political activity by voluntary associations will be directed largely toward narrowly defined objectives, such as neighborhood conditions or ethnic group rights. Neighborhood groups will be active and recognized by city officials as the spokesgroups for localistic interests. Economic interests—both of producers (business associations like the Chamber of Commerce) and of workers (unions)—will be more highly organized and politically active and influential than organizations stressing widely shared community concerns.

The level of political mobilization of mass memberships will be lower in pluralistic cultures, not only because of fewer middle-class residents but also because the subculture norms discourage mass participation as a means for affecting public policy. Organized political efforts by voluntary associations will be confined largely to activities by spokespersons presenting their clientele's views to city officials. In the absence of strong citywide uncontested voluntary associations capable of coordinating interorganizational cooperative action, this network is likely to be sparser, temporary, and less able to sustain common effort to affect policy.

Policy impacts of voluntary groups will also differ across cities embedded in different political subcultures. Public officials' responsiveness to organized interest group demands will vary, although the empirical evidence is inconsistent. For example, chief municipal executives in "reform" cities (presumably with more unitary cultures) evaluate interest groups, particularly middle-class groups, as more influential than do executives in "unreformed" cities (presumably more pluralistic cultures, see Northrop and Dutton, 1978). Reform structures make chief executives' career ambitions more vulnerable to a broad variety of group claims and interests. Hence, mayors and managers in such settings will be more sensitive and accommodating to interest group demands. But this analysis clashes with Lineberry and Fowler's (1967) wellknown finding that socioeconomic cleavages and ethnic heterogeneity (that is, pluralistic cultures) were less strongly related to taxation and expenditures in reformed cities than in cities not adopting reform institutions. Karnig

(1975) likewise found that stronger civil rights group mobilization promoted local private-regarding programs, but the group impact was blunted in reform cities. In summary, to the extent that the reform-nonreform distinction reflects significant political cultural variation across cities, the impact of interest groups on routine and special interest public policies remains empirically unresolved.

In the resolution of community controversy, voluntary associations may also differ in importance across political cultures. Coleman's (1957) theory of community polarization suggests that identification with the community can interrupt the process of conflict degeneration at the crucial point where disagreement turns to direct antagonism. Strong empirical support comes from Crain and Rosenthal's (1967) examination of urban renewal, school desegregation, and fluoridation decisions. While they did not measure political culture directly, they linked community socioeconomic status to difference in political values, assuming an aggregate relationship based on individual-level covariation between high-status and public-regarding attitudes toward government policies. Their second assumption was that high-status communities—presumably those with more unitary values—contain more and larger voluntary associations with greater capacity to mobilize people around issue-oriented campaigns, especially in reaction and opposition to controversial policy proposals. As a result, government officials will be more cautious and pressured to remain neutral; a controversial issue is likely to result in nothing being done. In contrast, in low-status communities (presumably those with more pluralistic cultures), citizens were less likely to be mobilized to influence decisions. When policy is made by a traditional machine or business-dominated elite, "there should be less controversy, and few programs, once begun, will be sidetracked" (Crain and Rosenthal, 1967).

Elite political cultural values may also affect voluntary associations' effectiveness in pressing claims on local government. Where elected and appointed officials view as illegitimate such efforts as petitioning for special interest programs, interest group inpact will be curtailed. The Bay Area study (Eulau and Prewitt, 1973) demonstrated substantial differences in city councils' receptivity to interest group influence efforts. The more favorably a council regarded group activity, the more likely the council was to facilitate group access (p. 336), a relationship persisting under a variety of statistical controls. Unfortunately, the study did not determine whether group activation and councils' attitudes toward groups affected the extent to which group policy objectives were implemented through council decisions.

In conclusion, scattered evidence implies that political subcultural values may facilitate or impede the operation and effectiveness of voluntary associations as agents of political influence. Much evidence is highly tangential and inferential. A thorough study with data collected for that primary purpose is required to pin down the extent to which political culture beliefs affect voluntary associations and other political system variables.

RESEARCH SUGGESTIONS

As should be apparent by now, the research evidence on political culture—especially American urban subcultures—is exceedingly sparse. Political norms and values, the "shoulds and oughts" of political life, simply have not received much empirical attention from political scientists and sociologists. This situation may be read alternately as benign neglect or an unheralded opportunity. Assuming the latter, this concluding section outlines some ways to turn that opportunity to advantage.

Several indirect indicators of political subculture of cities might be collected from census-type secondary sources. Population attributes unfortunately are ambiguous indicators and are not substitutes for direct measures of political norms and values. But for historical analysis, the population composition on immigration, ethnicity, racial, and religious indicators is virtually the only mass quantitative data available. Information on political institutions should be gathered, especially the dates for the adoption of various reform institutions, the creation of city planning and development departments, and the expenditures on amenities (for example, libraries and parks) provided by the city (see Knoke, forthcoming, for an example of the adoption of commission and city manager government). Of course, using such policy data to infer political culture confuses independent with dependent variables.

Content analyses of elite documents might be undertaken for qualitative clues to the dominant political values of a community. Such documents include newspaper editorials and letters to the editor, campaign speeches and inaugural addresses, party platforms, city charter preambles, and the like. This approach stands squarely in a tradition stretching back to Aristotle.

Elazar's (1970, 1972; Sharkansky, 1969) evaluations of the state and substate political cultural regions are available as global, if impressionistic, indicators of the larger subcultural contexts within which specific urban communities are situated. While applications of Elazar's mapping technique could probably be undertaken in each community in a small comparative study, the method is too elaborate and time-consuming to be adapted for a large number of cities.

At least two writers working in the Elazar tradition have explicitly acknowledged that survey data on individuals' political values and beliefs would be preferable—Monroe (1978) Sharkansky (1969). Setting aside the problem of sample size to focus first on content, we must recognize the absence of standardized political culture measures, with the possible exception of some indicators of political cynicism and distrust (usually with a national rather than a local target). The several handbooks of political and social measurement do not contain identifiable political culture scales—that is, items tapping general normative or value orientations toward political objects. Thus, any efforts to measure political subcultures must proceed largely from scratch. Creation of specific items will depend on which dimensions of political culture are most relevant for the substantive concerns of any research project the Chicago conference might launch. However, if pressed to limit content to a handful of items most likely to affect political system performance, I would stress norms about the scope of government and participation in political influence attempts as the major candidates for a mass electorate survey. Measuring political culture at the elite decision maker level is equally important as (if not more important than) measuring norms and values of the citizenry. As "proximate decision makers," elected officials' and bureaucrats' belief about the political process should be especially crucial in shaping policy decisions. Any questionnaires or interviews with urban elites should include a sizeable section on the basic value premises of these leaders.

To collect adequate information on community residents' political beliefs requires samples of considerable size if the responses are to be broken down by subculture. With samples on the order of 500 respondents per city, surveying more than a representative handful of communities rapidly becomes too costly. Neither does the main alternative—sampling design—offer much hope: National samples generally yield too few cases per primary sampling unit (at most about 25) to provide reliable estimates of city-level attitude distributions, let alone subgroup analyses. And efforts to simulate attitudes at lower levels of aggregation (Weber et al., 1972; Erikson, 1978) do not appear viable either. Thus, if individual political orientations are to be measured, the communities and neighborhoods from which samples are drawn must be carefully selected to maximize the representation of the full range of political cultures and their institutional manifestations.

A major presumption underlying much political culture writing is that norms and values about politics affect personal involvement in political life. Collection of extensive data from secondary sources (such as court records or newspapers) and direct survey interviews should be a top priority in any research on political culture. This is a list of only the most obvious topics:

aggregate registration and turnout levels in local elections;

party vote distributions by office;

party identification and membership;

party structure and competition;

party personnel pools and campaign effort;

voluntary association types and density;

level of membership and political orientation;

mobilization efforts by groups;

frequency of contacts with elected officials and bureaucrats and objectives sought;

types and frequency of protest incidents and levels of social movement activities.

I began my inquiry into the usefulness of the political culture for urban policy research with much skepticism, and I conclude with doubts remaining. The concept remains imprecise and defies easy measurement. The sparse literature, particularly empirical studies, testifies either to inherent operational difficulties or to lack of theoretical interest by political scientists and sociologists in studying the role of political norms and values in American political behavior, at either the individual or aggregate level.

But enough suggestive theoretical and empirical evidence exists to resist completely writing off the political culture idea as impractical. What is called for is a serious effort to develop precise quantitative indicators of basic political beliefs, preferably measured at the individual level, which can be incorporated into analyses of citizen and elite political action and decision-making. Only by giving the political culture concept a fair empirical test can we firmly decide whether it deserves to be rescued from metaphorical oblivion.

NOTES

1. Political ethos and its relationship to ethnicity and class has been the center of controversy almost since its inception. Hamilton (1975) analyzed 11 fiscal referenda surveys of

individuals and concluded that the "data furnish impressive support for the ethos theory," with occupational subculture possibly more important than ethnicity. Greeley and McCready (1974) found many political and social differences among Anglo-Saxon, Irish, and Italian Americans which they attributed to cultural traditions of the country of origin. Clark (1975) found a high correlation of Irish percentage and high municipal budgets, which he interpreted as an "Irish ethic of nonideological particularism" linked to a leadership style based on use of separable goods and resources allocated to persons and social groups.

The conclusion that affluent Americans are more "public-regarding" in support of issues not of direct personal benefit did not hold up in one study (Hoffman and Clark, 1978). Clark and Ferguson (forthcoming: 8-9) concluded that affluent Americans are becoming more fiscally conservative. They also suggest that the aggregate-level findings that affluent cities spend more than other cities may reflect a political coalition of the most and least affluent groups against the middle-income groups.

Miller and Bennett (1976) faulted the Wilson-Banfield study for confusing independent and dependent variables in "percentagizing" tables and for violating statistical criteria for index construction. Reanalysis, they argued, shows fewer sharp ethnic differences and greater importance of income and education. Vanecko and Kronenfeld (1977), while not directly measuring ethos, found patterns of ethnic-racial preferences for public spending in ten cities that were contrary to those predicted by Wilson and Banfield. Income effects seemed to be as strong as ethnic effects on public expenditure preferences. Their conclusion is "that political culture, largely a function of the social structure within ethnic groups, defines what is preferred, not on the basis of some political ethic, but rather in terms of a calculation of costs and benefits as seen across the experiences of relatives, friends, and acquaintances within one's self-defined community of shared self-interest and who is excluded." For other critiques of the Wilson-Banfield thesis, see Hennessey (1970), Hawkins and Prather (1971), and Durand (1972).

REFERENCES

AIGNER, S. (1977) "Political trust, political alienation and withdrawal from the political arena." Presented to American Sociological Association meetings, Chicago.

ALFORD, R. and H. SCOBLE (1969) Bureaucracy and Participation: Political Cultures in Four Wisconsin Cities. Chicago: Rand McNally.

ALMOND, G. and S. VERBA (1963) The Civic Culture: Political Attitudes and Democracy in Five Nations. Princeton: Princeton University Press.

BANFIELD, E. C. and J. Q. WILSON (1963) City Politics. New York: Vintage.

CLARK, T. N. (1975) "The Irish ethic and the spirit of patronage." Ethnicity 2 (December): 305-359.

——— and L. C. FERGUSON (forthcoming) "The middle class: policy preferences and political involvement," ch. 5 in Political Leadership and Urban Fiscal Policy.

COLEMAN, J. S. (1957) Community Conflict. New York: Free Press.

COULTER, P. B. (1970) "Comparative community politics and public policy." Polity 3 (Fall): 22-43.

CRAIN, R. L. and D. B. ROSENTHAL (1967) "Community status as a dimension of local decision making." American Sociological Review 32 (December): 970-985.

DEVINE, D. (1972) The Political Culture on the United States: The Influence of Member Values on Regime Maintenance. Boston: Little, Brown.

DOMHOFF, G. W. (1978) Who Really Rules? New Haven and Community Power Reexamined. Santa Monica, CA: Goodyear Publishing.

DURAND, R. (1972) "Ethnicity, 'public-regardingness,' and referenda voting." Midwest Journal of Political Science 16: 259-268.

DYE, T. R. (1976) Who's Running America? Englewood Cliffs, NJ: Prentice-Hall.

EASTON, D. (1965) A Framework for Political Analysis. Englewood Cliffs, NJ: Prentice-Hall.

ELAZAR, D. J. (1972) American Federalism: A View from the States. New York: Thomas Y. Crowell.

ELAZAR, D. J. and J. ZIKMUND [eds.] (1975) The Ecology of American Political Culture: Readings. New York: Thomas Y. Crowell.

——— (1970) Cities of the Prairie. New York: Basic Books.

ERIKSON, R. S. (1978) "Constituency opinion and congressional behavior: a reexamination of the Miller-Stokes representation data." American Journal of Political Science 22 (August): 511-535.

EULAU, H. and R. EYESTONE (1968) "Policy maps of city councils and policy outcomes: a developmental analysis." American Political Science Review 62 (March).

EULAU, H. and K. PREWITT (1973) Labyrinths of Democracy: Adaptations, Linkages, Representation and Policies in Urban Politics. Indianapolis: Bobbs-Merrill.

GREELEY, A. M. and W. C. McCREADY (1974) "Does ethnicity matter?" Ethnicity 1 (April): 91-108.

HAMILTON, H. D. (1975) "Political ethos; the evidence in referenda survey data." Ethnicity 2 (March): 81-98.

HARTZ, L. (1955) The Liberal Tradition in America. New York: Harcourt Brace Jovanovich.

HAWKINS, B. W. and J. E. PRATHER (1971) "Measuring components of the ethos theory: a first step." Journal of Politics 33: 642-658.

HENNESSY, T. M. (1970) "Problems in concept formation: the ethos 'theory' and the comparative study of urban politics." Midwest Journal of Political Science 14: 537-564.

HOFFMAN, W. L. and T. N. CLARK (1978) "Citizen preferences and urban policy types." Chicago: University of Chicago. (mimeo)

KARNIG, A. K. (1975) "'Private-regarding' policy, civil rights groups, and the mediating impact of municipal reforms." American Journal of Political Science 19 (February): 91-106.

KAVANAUGH, D. (1972) Political Culture. London: Macmillan.

KNOKE, D. (forthcoming) "The spread of municipal reform: spacial, temporal, and social dynamics." American Journal of Sociology.

LAUMANN, E. O. and F. U. PAPPI (1976) Networks of Collective Action: A Perspective on Community Influence Systems. New York: Academic Press.

LEHMAN, E. W. (1977) Political Society: A Macrosociology of Politics. New York: Columbia University Press.

LINEBERRY, R. L. and E. P. FOWLER (1967) "Reformism and public policies in American cities." American Political Science Review 66 (September): 701-716.

McCLOSKY, H. (1964) "Consensus and ideology in American politics." American Political Science Review 58 (June): 361-382.

MacPHERSON, M. (1978) "Voluntary affiliation: a structural approach." Charleston: University of South Carolina. (mimeo)
MERELMAN, R. M. (1969) "The development of political ideology: a framework for the analysis of political socialization." American Political Science Review 63 (September): 750-767.
MILLER, A. H. and S. E. BENNETT (1976) "Political ethos: some imperative empirical and conceptual reconsiderations." Ethnicity 3 (June): 99-123.
MILLER, A. H. (1974) "Political issues and trust in government: 1964-1970." American Political Science Review 68 (September): 951-972.
MONROE, A. D. (1978) "The cultural basis of American politics: testing Elazar's Theory." Presented to meetings of the Southern Political Science Association, Atlanta, Georgia.
——— (1975) Public Opinion in America. New York: Dodd, Mead.
NORTHROP, A. and W. H. DUTTON (1978) "Municipal reform and group influence." American Journal of Political Science 22 (August): 691-711.
O'CONNOR, E. (1956) The Last Hurrah. Boston: Little, Brown.
PHILLIPS, K. (1969) The Emerging Republican Majority. New Rochelle, NY: Arlington House.
PROTHRO, J. W. and C. W. GRIGG (1960) "Fundamental principles of democracy: bases of agreement and disagreement." Journal of Politics 22 (Spring): 276-294.
PUTNAM, R. D. (1971) "Studying elite political culture: the case of 'ideology'." American Political Science Review 65 (September): 651-681.
PYE, L. (1968) "Political culture," pp. 218-225 in D. L. Sills (ed.) International Encyclopedia of the Social Sciences, Vol. VIII. New York: Macmillan.
RITT, L. G. (1974) "Political cultures and political reform: a research note." Publius 4: 127-133.
ROSENBAUM, W. A. (1975) Political Culture. New York: Praeger.
SAVAGE, R. L. and R. J. GALLAGHER (1977) "Politicocultural regions in a Southern state: an empirical typology of Arkansas counties." Publius 7: 91-105.
SHARKANSKY, I. (1969) "The utility of Elazar's political culture: a research note." Polity 2 (Fall): 66-83.
STOUFFER, S. (1955) Communism, Conformity, and Civil Liberties. Gloucester, MA: Peter Smith.
TURK, H. (1977) Organizations in Modern Life. San Francisco: Jossey-Bass.
VANECKO, J. J. and J. KRONENFELD (1977) "Preferences for public expenditures and ethno-racial group membership: a test of the theory of political ethos." Ethnicity 4 (December): 311-336.
VERBA, S. and N. H. NIE (1972) Participation in America: Political Democracy and Social Equality. New York: Harper & Row.
WEBER, R. E. and W. R. SHAFFER (1972) "Public opinion and American state policy-making." Midwest Journal of Political Science 16: 683-699.
WEBER, R. E., A. H. HOPKINS, M. L. MEZER, and F. J. MUNGER (1972) "Computer simulation of state electorates." Public Opinion Quarterly 36 (Spring): 49-65.
WEISSBERG, R. (1976) Public Opinion and Popular Government. Englewood Cliffs, NJ: Prentice-Hall.
WILLIAMS, O. (1961) "Typology for comparative local government." Midwest Journal of Political Science 5 (May): 150-164.
WILSON, J. Q. [ed.] (1968) City Politics and Public Policy. New York: John Wiley.
——— and E. C. BANFIELD (1971) "Political ethos revisited." American Political Science

Review 65 (December): 1048-1062.
——— (1964) "Public-regardingness as a value premise in voting behavior." American Political Science Review 58 (December): 876-887.
WRIGHT, J. D. (1976) Dissent of the Governed: Alienation and Democracy in America. New York: Academic Press.

10

Citizen Preferences and Policy Responsiveness

PAUL D. SCHUMAKER

SINCE THE LANDMARK STUDY of Lineberry and Fowler (1967) into the relationship between municipal governmental structure and responsiveness, substantial attention has been given to investigating the responsiveness of city governments to citizen preferences. The central questions addressed in these studies include: (1) How responsive are local policymakers (Parenti, 1970)? (2) Which types of citizens (for example, active and inactive, blacks and whites) are most likely to have their preferences receive a favorable response from policymakers (Verba and Nie, 1972; Schumaker and Getter, 1977)? (3) On what kinds of issues are policymakers most responsive (Sutton, 1973; Schumaker and Loomis, 1979)? (4) By what processes are citizen preferences converted to responsive policies (Luttbeg, 1974)? (5) What are the barriers to responsive local government (Zeigler and Tucker, 1978; Weber, 1975)? (6) What types of citizen behaviors enhance responsiveness (Verba and Nie, 1972; Schumaker and Billeaux, 1978)? (7) What political, economic, and social characteristics of cities affect responsiveness (Eulau and Prewitt, 1973; Karnig, 1975; Getter and Schumaker, 1978a)? Although there is apparent agreement on the theoretical and practical importance of these questions, there is little consensus among urban scholars about their answers or about how to study them.

AUTHOR'S NOTE: *This research was supported by the General Research Fund of the University of Kansas. My understanding of the ten issues presented in this chapter has profited from discussion and work with my colleague, Russell Getter. His numerous contributions are acknowledged and appreciated, though he bears no responsibility for the conclusions presented here.*

In this chapter ten unresolved issues which have retarded cumulative research on responsiveness are considered. By focusing on these issues, it is possible to review and assess various approaches to the study of the responsiveness of local governments. The principal assumption underlying these assessments is that the purpose of focusing on responsiveness is to develop descriptive and explanatory theory, supported by convincing empirical evidence, concerning the linkage between the policy preferences of citizens of political communities and the public policies of their governments. It is assumed that the link between citizen preferences and governmental policies is always imperfect; various degrees of "translation error" (Godwin and Shepard, 1976) occur depending on the characteristics of the people involved and the organizational, political, social, and economic settings in which politics occurs.

When discussing these issues, attention will be focused on responsiveness of municipal governments to holistic citizen preferences. This focus is for purposes of brevity only. Clearly, the central questions about responsiveness concern how and why officials respond differentially to the diverse preferences of various subpopulations within communities. Thus, while the discussion emphasizes responsiveness to the public generally, it also pertains to questions about responsiveness to white citizens, to black citizens, to upper-status citizens, to lower-status citizens, and so on.[1] And while the discussion emphasizes responsiveness of municipal governments, it also pertains to other governmental bodies (such as school districts and county governments) in urban settings.

Some very broad recommendations for the future study of responsiveness result from analysis of the ten issues. The major conclusion is that large-scale cross-community investigations are currently feasible only through the utilization of crude surrogate measures of important components of policy responsiveness. Such studies may have important heuristic functions, but they cannot provide convincing evidence confirming central theoretical propositions. More rapid development and validation of theory on responsiveness may occur by conducting case studies using a common theoretical framework and methodological approach. While it is possible to indicate the broad outlines of such an approach, a number of problems must still be solved to make it operational.

THE CONCEPT OF RESPONSIVENESS

Like democracy and justice, responsive government is praised by almost everyone, though there is little consensus about the meaning of the term. In this section, three issues involving the definition and measurement of responsiveness are considered.

Issue 1: In order to be considered responsive, to what extent must urban policymakers act according to citizen preferences?

According to Pitkin (1972: 209-210), political representation requires that policymakers "act in the interests of the represented, in a manner responsive to them . . . that the representative must not be found persistently at odds with the wishes of the represented without good reason in terms of their interest, without a good explanation of why their views are not in accord with their interests." This formulation, which is the starting point of most recent studies of responsiveness, is ambiguous about the extent to which responsive policymakers can act contrary to the will of the public.

Most researchers have emphasized the norm of not being "at odds with the wishes" of citizens. By defining and measuring responsiveness as the extent to which policies are congruent with citizen policy preferences, these researchers categorize as "unresponsive" any policy actions which are inconsistent with dominant preferences, regardless of whether these actions are justified by policymakers as meeting the real interests of constituents.

In contrast, Eulau and Karps (1978: 69) argue that policymakers need merely "stand ready to be responsive"—what Pitkin emphasized is "a potential for response rather than the act of response." Thus, in this approach, the proper indicators of whether or not policymakers are responsive are their role orientations or "representational response styles." If policymakers indicate in interviews or questionnaires that they consider the views of attentive publics or issue groups, they are categorized as responsive; those who claim to be immune from external pressures, acting instead on the basis of "self-defined images(s) of what community needs are," are labeled unresponsive (Prewitt and Eulau, 1969: 430).

Both of these formulations have obvious merit. If the purposes of studying responsiveness are to develop a normative theory of representative government or a standard for evaluating the performance of local policymakers, then the conception of responsiveness as congruence between citizen preferences and public policy may be inadequate. It does an injustice to the richness of Pitkin's notion of representative government. It is neither possible nor desirable for policymakers always to act in accordance with dominant citizen preferences. But if the purpose of studying responsiveness is to develop an empirical theory of the translation error which occurs in converting citizen inputs into policy, then responsiveness should be conceptualized as preference-policy congruence. As suggested by Miller and Stokes (1963), there are two components to translation error: (1) the gap between actual citizen preferences and citizen preferences as perceived by policymakers, and (2) the gap between perceived citizen preferences and public policy. Both are identifiable using the congruence approach, as actual preferences can be related to perceived preferences and perceived preferences can be related to policy actions (Schumaker et al., 1979: 26). But studies of representational response styles tell us nothing about either of these gaps. The mere fact that policymakers indicate they are willing to listen to citizens and groups provides little assurance that they will accurately perceive public wants. An adequate theory of responsiveness must consider the extent to which policymakers misperceive citizen preferences and the role political institutions play in bringing about these misperceptions. And even if policymakers listen to and accurately perceive citizen wants, a responsive predisposition by officials does not ensure that their actual policy-making behavior will reflect citizen preferences. Thus, empirical theory about political linkages requires that responsiveness be conceptualized as congruence between citizen wants and public policy.

Issue 2: Does responsiveness connote a unidirectional relationship in which citizen preferences cuase public policy, or does it connote mere correlation between citizen preferences and policy?

The relationship between citizen preferences and policy actions has two component parts: the influence of citizens on policymakers and the influence of policymakers on citizens (Edelman, 1971; Erikson, 1976: 28-30; Shaffer and Weber, 1974; Schumaker and Loomis, 1979). Because many normative democratic theorists (Pateman, 1970; Bachrach, 1967) hold that policymakers should respond to citizen initiatives, it has been questioned whether responsiveness is also achieved when policymakers influence public attitudes, bringing them into alignment with policy decisions.

There are several reasons for viewing responsiveness as preference-policy congruence without specifying how congruence is achieved. First, philosophical conceptions of political representation seldom suggest that policymakers should be merely reactive. As Eulau and Karps (1978: 69) point out, representatives have a "superior position" vis-à-vis citizens because they are chosen to make decisions. Policymakers should thus exercise political leadership, proposing programs and policies responsive to community needs (Burns, 1978). When such initiatives are supported by citizens, policymakers should not be labeled unresponsive merely because the initiative does not come from the people. Second, citizens seem better able to generate meaningful preferences in response to policymaker behavior than to articulate policy initiatives (Wahlke, 1978). Responsiveness as congruence recognizes that the requisites of representative democracy have been fulfilled if policymakers can maintain support for their activities. Third, disentangling the causal processes in the citizen-policymaker relationship is an important but complex methodological problem which need not be solved if responsiveness is conceptualized as congruence. Thus, while citizen influence over policy involves a unidirectional causal relationship flowing from citizens to policymakers, responsiveness should be viewed as congruence between citizens and policy without regard for the causal processes involved (Schumaker et al., 1979: 21-23).

Issue 3: Should responsiveness be analyzed principally as a dyadic or collective relationship between citizens and policymakers?

In the literature on state legislatures and congress, responsiveness is often viewed as a dyadic relationship of legislators and their districts (Kuklinski, 1978; Weissberg, 1978). However, studies of local responsiveness have normally adopted a collective perspective on representation. While dyadic representation concerns whether *individual* representatives respond positively to dominant citizen preferences in their districts, collective representation concerns whether *collective* decisions of a policy-making body are consistent with dominant preferences of all citizens in the political community. Weissberg (1978: 536-537) has argued that collective representation is the more useful concept because, unlike dyadic representation, it recognizes the importance of virtual representation, which occurs when citizen preferences are represented by policymakers other than their district representative. Since many American cities do not have ward structures which allow for dyadic representation, studying responsiveness of local governments from the perspective of collective representation is especially desirable.[2]

PUBLIC POLICY

Issue 4: What types of policy actions by urban officials should be considered to determine their responsiveness?

Eulau and Karps (1978: 62-67) have argued that "there are four possible policy-related components of responsiveness which, as a whole, constitute representation": policy responsiveness, service responsiveness, allocation responsiveness, and symbolic responsiveness. Policy responsiveness is concerned with policymakers being attentive to citizen preferences on "the great issues that agitate the political process." Service responsiveness involves bestowing benefits and advantages on particular constituents who request them. Allocation responsiveness is concerned with pork barrel allocations of goods and services to specific spatial locations and has two distinct components in urban politics. One dimension concerns local officials seeking and receiving intergovernmental discretionary funds for community programs, projects, and facilities (Aiken and Alford, 1970; Clark, 1968). A second dimension concerns the intracity cross-neighborhood allocation of public services; when policymakers concentrate parks, police or fire services, or other public benefits in specific neighborhoods, they are exhibiting allocation responsiveness to the residents of that neighborhood (Levy et al., 1974; Lineberry, 1977, 1978). Finally, symbolic responsiveness is concerned with those "significant symbolic gestures . . . [by policymakers which] . . . generate and maintain continuing support of citizens."

Each of these is an important element of urban policy-making. Nevertheless, policy responsiveness is the concept that should receive the most attention by those interested in political linkage and translation error. Service responsiveness is peripheral to these concerns because such activities usually have no significant impact on public policy, at least when public policy is defined as those governmental activities having broad and persistent impacts on the public (Eulau and Prewitt, 1973: 473). The two dimensions of allocation responsiveness are also peripheral to these central concerns because they deal with policymaker behaviors without considering citizen preferences. Urban policymakers are not responsive to citizen preferences when they obtain a federal grant for a redevelopment project opposed by many citizens. And urban administrators are not equally responsive to preferences of lower-status and upper-status citizens when higher quality streets in upper-status neighborhoods are offset by expensive expressways in lower-status neighborhoods. Symbolic responsiveness is also a concept analyzed without adequate consideration of citizen preferences. If citizens prefer tangible policies instead of symbolic gestures, and if symbolic gestures fail to satisfy citizen concerns, it is difficult to see how policymakers who depend on symbols are responsive. Thus, only policy responsiveness deals with the linkage between citizen preferences and policy.

Issue 5: What efforts must policymakers make at various stages of the policy-making process in order to be responsive?

In their important study, Verba and Nie (1972: 300) suggested that "leaders are responsive if they accurately perceive citizen priorities, if they agree with the citizenry on the nature of community problems, and if they are active in trying to solve those problems." While this formulation asks only that policymakers share the same policy agenda as citizens, most students of responsiveness have demanded more policy-making behavior. For some, the test is whether policy enactments are consistent with citizen preferences (Shaffer and Weber, 1974). For others, enacted policies should be fully implemented and/or enforced (Schumaker and Loomis, 1979).

These various approaches suggest that public officials make a variety of efforts to be responsive; they make themselves accessible to citizens; they hold hearings; they enact ordinances; they enforce laws; and they deliver services. In each such activity they may or may not behave according to citizen preferences. Thus to account for potential discrepancies in responsiveness of policymakers at different stages in the policy process, the level of policy effort by urban officials should be considered in each of three stages: (1) the institutional (agenda) stage, concerning the extent to which officials actively consider issues and weigh various courses of action or inaction; (2) the enactment stage, concerning the extent to which laws are made; and (3) the implementation stage, concerning the extent to which enacted laws are implemented and/or enforced (Getter and Schumaker, 1978b; Cobb and Elder, 1972).[3]

CITIZEN PREFERENCES

Issue 6: What citizen inputs should be analyzed to determine the level of responsiveness of urban policymakers?

The principal issue raised by this question is whether responsive policymakers consider only preferences of those citizens who communicate with them or if they somehow consider preferences of all citizens, including those who fail to articulate and communicate their desires (Verba and Nie, 1972: 268). Research has focused on responsiveness to specific group demands and individual requests (Karnig, 1975; Zeigler et al., 1974; Schumaker and Billeaux, 1978), responsiveness to public opinion (Verba and Nie, 1972; Hansen, 1975; Schumaker and Getter, 1977), and responsiveness to both group demands and public opinion (Getter and Schumaker, 1978a). A focus on responsiveness to public opinion provides the broadest understanding of political linkage. Studies of articulated demands capture mainly responsiveness to active, disproportionately upper-status citizens (Zeigler and Tucker, 1978: 5-6). But studies which fail to consider responsiveness to group demands and individual requests may not indicate to whom policymakers are principally responsive (Lowi, 1979: 167-197) and thus will fail to suggest why higher levels of responsiveness to public opinion are not achieved (Weber and Shaffer, 1972). Responsiveness to both public opinion and group demands should thus be studied.

Issue 7: How are articulated demands to be measured?

Investigations of responsiveness to group demands have been plagued by difficulties in measuring articulated demands. Crude indicators of group organization (for example, membership in the NAACP by community) have been used in comparative urban research as surrogate indicators of group demands (Karnig, 1975). And case studies have focused on whether designated groups have received what they requested, without adequate consideration of counterpressures by other groups (Lipsky, 1970; Parenti, 1970). To move beyond these difficulties, several problems must be considered.

First is the operational definition of "demand" (Greenstone, 1975: 244). For example, it is often difficult to distinguish citizen communications seeking clarification of and/or justification for existing policies from communications indicating clear policy preferences. Second, because policymakers usually receive communications supporting conflicting policy positions, both support of and opposition to policy proposals must be included in the measure of "net demands." But this raises a third problem: In calculating "net demands," how are various communications to be weighed? Since there is little justification for considering a single telephone call from one individual as an equivalent demand to persistent communications from leaders of large and powerful organizations, it makes poor sense to treat all demands equally. To deal with this problem, group demands should be weighed by the amount of "pressure"—an index involving group resources and activities—which accompanies each demand (Getter and Schumaker, 1978b). However, research has not been conducted establishing a valid index of group pressure.

Even if these conceptual problems are solved, data problems remain. It would be desirable to gather information directly from a random sample of active groups and citizens regarding their policy goals, activities, and resources. But because little is known about the universe of citizen organizations concerned with local politics (Newton, 1976: 42), it is unclear how to sample active communications. Perhaps the most feasible strategy currently available is to narrow studies to a few issues (see issue 9 below) and use a type of network analysis which identifies, through interviews with public officials, those persons and groups who communicate their concerns and which then permits, using snowball sampling techniques, successive interviews with activists and group leaders themselves (Bolland, 1979). Yet, the feasibility of this strategy is inversely proportional to the number of issues and communities under investigation.

Issue 8: How should unarticulated citizen preferences be measured? For intracommunity studies of responsiveness, survey research is the obvious method for measuring public opinion. Random digit dialing enables easy and inexpensive representative samples to be drawn (Cummings, 1979). Policy preference questions can be crafted to fit specific issues appropriate to the community under investigation. Questions tapping the awareness of respondents about these issues can filter out "non-attitudes" (Converse, 1970). Although technical problems with survey instruments should not be minimized (Clark, 1976), it is clear that survey measures of citizen preferences can be related to public policy measures to obtain indices of the level and distribution of responsiveness within communities.

In cross-community research to date, however, the high costs of administering equivalent surveys in a large sample of cities have resulted in use of surrogate measures of citizen preferences. Four are considered briefly.

The crudest surrogate indicators of citizen preferences are census data concerning socioeconomic characteristics of city residents (Lineberry and Fowler, 1967; Godwin and Shepard, 1976). Although microlevel research suggests that socioeconomic variables are reasonable predictors of individual preferences (Fowler, 1974; Loverich, 1974), this strategy must be viewed cautiously for several reasons. First, it is unclear whether socioeconomic variables measure "preferences" or "need" (Sharkansky, 1971). Second, socioeconomic variables may be good predictors of citizen preferences in some policy areas but not in others (Dawson, 1973). Third, while preferences may be a function of socioeconomic variables, they are also a function of contextual variables (for example, local press coverage or specific community problems) not measured by socioeconomic characteristics (Brunner, 1977). Fourth, in analyses with socioeconomic variables as indicators of citizen preferences, responsiveness is measured by the degree of correlation across communities between socioeconomic characteristics of cities as independent variables and measures of public policy as dependent variables (Lineberry and Fowler, 1967). This strategy is questionable on theoretical grounds because responsiveness is a community-level property which should be measured on the basis of intracommunity political activity (Getter and Schumaker, 1978b). Thus socioeconomic characteristics are not adequate indicators of citizen references.

Referenda results provide another method of measuring citizen preferences in cross-community research (Erikson et al., 1975; Kulinski, 1978). But they raise two problems. First, referenda on equivalent issues should be analyzed in each city, but such data are not normally available in a national sample. Second, because the referenda voter tends to be disproportionately upper-status (Clubb and Traugott, 1977: 145), the preferences measured by referenda may not be representative of all citizens. Thus the usefulness of referenda data as measures of community policy preferences is limited.

A third way to estimate citizen preferences is to measure elite perceptions about the public (Clark and Ferguson, 1981: 18-19). As suggested in the discussion of issue 1, the problem with this strategy is that there are significant gaps between actual citizen preferences and elite perceptions of these preferences (Hedlund and Friesema, 1972; Miller and Stokes, 1963). Thus, if elite perceptions are to be useful as indicators of citizen preferences, research must be conducted to identify the characteristics of those elites most accurate in perceiving public concerns.

A fourth method for estimating citizen preferences involves simulation models (de Sola Pool et al., 1965; Weber, 1971; Schumaker, 1977). Simulation models first generate predictions regarding policy preferences of different "citizen types" living within communities, and then calculate community-level policy preference scores after weighing the preferences of different citizen types by their proportion in the community. Existing simulation models have been criticized on many grounds. The most important limitation of models currently in use is their assumption that citizen preferences are solely a function of selected demographic variables (Seidman, 1975: 335). Effects of community context on preferences are ignored. A second important limitation is that the model used assumes linear, additive relationships among demographic variables and citizen preferences, an assumption challenged on theoretical (Seidman, 1975) and empirical grounds (Kuklinski, 1977). Despite these limitations, however, it has been shown that simulation models can yield reasonably accurate estimates of citizen priorities among various municipal services (Schumaker, 1977).

In summary, simulation models probably provide the most valid and feasible method for estimating citizen preferences in a large number of cities for responsiveness studies. However, existing models have their limitations and are poor substitutes for survey research. Thus, the most valid measures of citizen preferences can be obtained only through survey responses in intracommunity case studies.

POLITICAL ISSUES

Issue 9: What political issues should be analyzed to determine the policy responsiveness of local governments?

Any community includes a huge number of issues. Some are of concern to many people, while others concern only a few. Some reach the institutional agenda; others do not.[4] Some are very abstract (for example, reducing crime), while others are concrete (such as whether or not to increase police surveillance in a particular neighborhood). As suggested by Polsby, "there seem to be no satisfactory criteria which would identify a universe of decisions [issues] in a community so that a sample of 'typical' or 'representative' decisions could in principle be drawn" (Frey, 1971: 1064). Given these difficulties, researchers have normally selected issues based on their political and/or scholarly interests or availability of data. Without more defensible criteria, it is difficult to assess the generality of findings from such studies.

One suggestion has been to focus on the "great issues which agitate the political process" (Eulau and Karps, 1978: 62). However, the criticisms by Bachrach and Baratz (1962) of the selection of key issues in the study of community power are equally relevant to the study of responsiveness. The analysis of only controversial issues is likely to result in distorted estimates of the overall responsiveness of policymakers, as evidence suggests that policymakers are most responsive on such issues (Miller and Stokes, 1963; Kuklinski, 1978; Schumaker and Loomis, 1979).

To avoid the problems involved in selecting only "great" or "key" issues, representative samples of key issues, routine issues, and nonissues would all have to be analyzed. However, because few citizens are likely to have well-informed preferences on routine issues (Zeigler and Tucker, 1978: 212) and because of difficulties in selecting nonissues (Wolfinger, 1971), this strategy seems foreclosed.

One possible approach to these difficulties is to develop "policy concepts" (Clausen, 1973: 16-17; Kuklinski, 1978) grounded in political theory and philosophy which suggest fundamental types of issues which all political communities must address. For example: (1) What should be the level of spending and taxation by local governments? (2) How should governmental resources be distributed among various functions or services? For example, should resources go to "people-helping" or "community-serving" functions (Wilson and Banfield, 1971: 1052)? (3) To what extent should property and industry be regulated, or to what extent should property rights be preserved? (4) To what extent should local government subsidize or benefit specific individuals and industries in pursuit of collective values, such as enhancing the employment base or preserving downtown business districts? (5) Should tax obligations be allocated progressively, proportionately, or regressively? (6) Should governmental benefits be distributed on a compensatory basis to the most needy, on a basis of equity where benefits are proportionate to contribution, or on the basis of equality? (7) To what extent should behavior contrary to community norms be regulated and subjected to criminal sanction?

Having developed a list of such fundamental issues, specific examples of these issues relevant to the community under investigation can be selected. The criteria for the selection of issues in communities are thus of two kinds. First fundamental issues are selected on the basis of their significance in political theory and philosophy. Second, specific issues are selected for their relevance to the community under study and for their "degree of fit" with the chosen fundamental issues. Selecting issues on this basis has several advantages. First, although the criteria involve researcher judgments, these can be defended on nonarbitrary grounds. Second, the criteria provide a basis for selecting nonissues. For example, substantial inequalities in property tax assessments may not be a controversial issue in a community—and may even be suppressed as an issue—but the closeness of its fit to the fundamental tax obligation issue (number 5) and its probable relevance to the community suggest its analysis. Third, although citizens may lack information and informed preferences about some specific community issues, they are likely to have meaningful and measurable preferences on the fundamental ones. Preferences on fundamental issues may be reliable predictors of preferences on specific issues and used accordingly in the analysis of governmental responsiveness (Nisbet, 1975).

COMPARATIVE ANALYSIS AND CASE STUDIES

Issue 10: Should more attention be given to comparative urban (cross-community) research or to case studies as methods for studying responsiveness at the local level?

Research on community power, the conceptual parent of policy responsiveness (Schumaker et al., 1979: 20-25), began by conducting a large number of case studies. Because these studies were motivated by diverse ideological perspectives and were conducted using different and controversial research methods, they did not produce cumulative findings (Aiken, 1970). Perhaps as a consequence, much of the research on the responsiveness of local governments has been conducted using comparative analysis rather than case studies. But, as suggested by the discussion thus far, comparative research has not always produced cumulative findings, owing to diverse conceptual frameworks, arbitrary selection of issues, and inadequate measures of citizen preferences, group demands, and public policy.

In contrast, slow but steady progress can be made through case studies which avoid ideological, conceptual, and methodological quarrels. By conducting a variety of case studies using equivalent theoretical frameworks and comparable data collection procedures, the basis for future comparisons across communities can emerge. Or, by developing a deductive theory of responsiveness which predicts specific patterns of responsiveness in "crucial cases," there may be little need to validate theory using inductive procedures and a large sample of communities (Eckstein, 1975; Schumaker, 1981). In short, case studies may provide the most feasible basis, at present, for developing a theory concerning the responsiveness of local governments to citizen preferences.

CONCLUSIONS

The following recommendations regarding the study of responsiveness emerge from considering the ten issues presented above. First, the patterns of responsiveness in specific communities, conceived as the degree of linkage and translation error between the preferences of various citizens and the actions of policymakers, should be investigated through case studies. Second, such studies should begin by selecting specific community issues because they represent fundamental philosophical issues confronting all communities. Controversial issues, routine issues, and nonissues should all be sampled. Third, public opinion on both the fundamental issues and specific manifestations of these issues should be measured through survey research. Samples should be large enough to allow disaggregation into various subpopulations (for example, blacks and whites) to enable investigations of possible differential or unequal responsiveness. Fourth, group demands, activities, and resources pertaining to the specific issues under investigation should be measured by interviews with policymakers, activists, and group leaders. Indices of "group pressure" in support of and in opposition to various policies should be developed. Fifth, for the issues under investigation, the policy efforts of governmental officials should be measured at the institutional, enactment, and implementation stages of the policy process. Sixth, the pattern of responsiveness in communities should be determined by relating these policy efforts to the preferences of citizens taken as a whole, to the preferences of various subpopulations, and to the demands of various types of groups. Seventh, the resulting empirical descriptions of responsiveness should be compared to theoretical expectations and previously completed case study findings to enhance cumulative theory and research.

Even if agreement on these points is reached, many problems remain. For example, what specific measurement instruments should be used in ascertaining citizen preferences? How are various group activities and resources to be weighed in order to develop indices of group pressure? How are policy efforts to be measured at various stages of the policy process? How are measures of citizen inputs and policy outputs to be analyzed to generate measures of responsiveness? And, perhaps most important, what is the prevailing theory for deriving 'theoretical expectations" against which to compare research results? Important as these questions are, there seems little need to address them until agreement is reached on the ten issues discussed here from which they are derived.

NOTES

1. The concepts of differential responsiveness (Verba and Nie, 1972: 315-318) and responsiveness bias (Schumaker and Getter, 1977) permit assessment of the extent to which policymakers are more responsive to one subpopulation (for example, whites) than to its categorical counterpart.

2. This is not to suggest that it would be inappropriate to study how different sets of urban policymakers vary in their responses to citizen inputs. As Rossi et al. (1974) suggest, mayors, councilmembers, and heads of municipal agencies can display different responsiveness patterns.

3. Different measurement procedures must be used to determine policy effort at each stage. For example, the extent to which policymakers consider policy proposals at the institutional stage can probably be measured best through interviews with policymakers and group leaders. Policy effort at the enactment stage can be measured using documentary data (such as minutes of council meetings, ordinances, and administrative correspondence). And policy effort at the implementation stage may use both interview and documentary data (for example, of changes in personnel and resource allocations).

4. Some issues will not make it to the institutional agenda of municipal governments simply because other governmental units have jurisdiction in the area. Obviously, studies of the responsiveness of municipal government should consider only those issues within their jurisdiction.

REFERENCES

AIKEN, M. (1970) "The distribution of community power: structural bases and social consequences," pp. 487-525 in M. Aiken and P. Mott (eds.) The Structure of Community Power. New York: Random House.

——— and R. R. ALFORD (1970) "Community structure and innovation: the case of public housing." American Political Science Review 64 (3): 843-864.

BACHRACH, P. (1967) The Theory of Democratic Elitism. Boston: Little, Brown.

——— and BARATZ, M. (1962) "The two faces of power." American Political Science Review 62 (4): 947-952.

BOLLAND, J. M. (1979) "Conflict and consensus in American communities: agenda-setting as prelude to social change." Ph.D. dissertation, Ohio State University. (unpublished)

BRUNNER, R. D. (1977) "An intentional alternative in public opinion research." American Journal of Political Science 21 (3): 435-464.

BURNS, J. M. (1978) Leadership. New York: Harper & Row.

CLARK, T. N. [ed.] (1976) Citizen Preferences and Urban Public Policy: Models, Measures, and Uses. Beverly Hills, CA: Sage

——— (1968) "Community structure, decision-making, budget expenditures, and urban renewal in 51 American communities." American Sociological Review 33(3): 576-592.

——— and L. C. FERGUSON (1981) Political Process and Urban Fiscal Strain, (unpublished manuscript)

CLAUSEN, A. R. (1973) How Congressmen Decide. New York: St. Martins' Press.

CLUBB, J. M. and M. W. TRAUGOTT (1972) "National patterns of referenda voting," pp. 137-169 in H. Hahn (ed.) People and Politics in Urban Society. Beverly Hills, CA: Sage.

COBB, R. W. and C. D. ELDER (1972) Participation in American Politics: The Dynamics of Agenda-Building. Boston: Allyn & Bacon.

CONVERSE, P. E. (1970) "Attitudes and non-attitudes: continuation of a dialogue," in E. Tufte (ed.) Quantitative Analysis of Social Problems. Reading, MA: Addison-Wesley.

CUMMINGS, K. M. (1979) "Random digit dialing: a sampling technique for telephone surveys." Public Opinion Quarterly 43 (2): 233-244.

DAWSON, R. A. (1973) Public Opinion and Contemporary Disarray. New York: Harper & Row.

de SOLA POOL, I., R. P. ABLESON, and S. POPKIN (1965) Candidates, Issues and Strategies: A Computer Simulation of the 1960 and 1964 Presidential Elections. Cambridge: MIT Press.

ECKSTEIN, H. (1975) "Case study and theory in political science." pp. 79-138 in F. I. Greenstein and N. W. Polsby (eds.) Handbook of Political Science., Vol. 7. Reading, MA: Addison-Wesley.

EDELMAN, M. (1971) Politics as Symbolic Action. Chicago: Markham.

ERIKSON, R. S. (1976) "The relationship between public opinion and state policy: a new look based on some forgotten data." American Journal of Political Science 20 (1): 25-36.

——— N. R. LUTTBEG, and W. V. HOLLOWAY (1975) "Knowing one's district: how legislators predict referendum voting." American Journal of Political Science 19 (2): 231-246.

EULAU, H. and P. D. KARPS (1978) "The puzzle of representation: specifying components of responsiveness," pp. 55-72 in H. Eulau and J. Wahlke (eds.) The Politics of Representation. Beverly Hills, CA: Sage.

EULAU, H. and K. PREWITT (1973) Labyrinths of Democracy: Adoptions, Linkages, Representation, and Policies in Urban Politics. Indianapolis: Bobbs-Merrill.

FOWLER, F. J., Jr. (1974) Citizen Attitudes Toward Local Government, Services, and Taxes. Cambridge, MA: Ballinger.

FREY, F. W. (1971) "Comment: on issues and non-issues in the study of power." American Political Science Review 65 (4): 1081-1101.

GETTER, R. W. and P. D. SCHUMAKER (1978a) "Contextual bases of responsiveness to citizen preferences and group demands." Policy and Politics 6 (1): 249-278.

——— (1978b) "Responsiveness and responsiveness bias: a conceptual and measurement agenda for intra-community analyses and comparative research." Presented at the Midwest Political Science Association, Chicago, 3, April 20-22.

GODWIN, R. K. and W. B. SHEPARD (1976) "Political processes and public expenditures: a re-examination based on theories of representative government." American Political Science Review 70 (4): 1127-1135.

GREENSTONE, J. D. (1975) "Group theories," pp. 243-318 in F. I. Greenstein and N. W. Polsby (eds.) Handbook of Political Science, Vol. 7. Reading, MA: Addison-Wesley.

HANSEN, S. B. (1975) "Participation, political structure, and concurrence." American Political Science Review 69 (4): 1181-1199.

HEDLUND, R. P. and H. P. FRIESEMA (1972) "Representatives' perceptions of constituency opinion." Journal of Politics 34 (3): 730-752.

HOFFMAN, W. L. and T. N. CLARK (1979) "Citizen preferences and urban policy types," pp. 85-106 in J. P. Blair and D. Nachmias (eds.) Fiscal Retrenchment and Urban Policy. Beverly Hills, CA: Sage.

KARNIG, A. K. (1975) " 'Private-regarding' policy, civil rights groups, and the mediating impact of municipal reforms." American Journal of Political Science 19 (1): 91-106.

KUKLINSKI, J. H. (1978) "Representativeness and elections: a policy analysis." American Political Science Review 72 (1): 165:177.

——— (1977) "Constituency opinion: a test of the surrogate model." Public Opinion Quarterly 41 (1): 34-40.

LEVY, F., A. J. MELTSNER, and A. B. WILDAVSKY (1974) Urban Outcomes. Berkeley: University of California Press.

LINEBERRY, R. L. (1978) The Politics and Economics of Urban Services. Beverly Hills, CA: Sage.

——— (1977) Equality and Urban Policy: The Distribution of Municipal Public Services. Beverly Hills, CA: Sage.

——— and FOWLER, E. (1967) "Reformism and public policies in American cities." American Political Science Review 61 (3): 701-716.

LIPSKY, M. (1970) Protest in City Politics. Chicago: Rand McNally.

LOVERICH, N. P., Jr. (1974) "Differing priorities in an urban electorate: service preferences among Anglo, Black, and Mexican-American voters." Social Science Quarterly 55 (4): 704-717.

LOWI, T. J. (1979) The End of Liberalism. New York: W. W. Norton.

LUTTBEG, N. R. (1974) Public Opinion and Public Policy. Homewood, IL: Dorsey Press.

MILLER, W. E. and D. E. STOKES (1963) "Constituency influence in congress." American Political Science Review 57 (1): 45-56.

NEWTON, K. (1976) "Feeble government and private power: urban politics and policies in the United States," pp. 37-58 in L. H. Masotti and R. L. Lineberry (eds.) The New Urban Politics. Cambridge, MA: Ballinger.

NISBET, R. (1975) "Public opinion and popular opinion." The Public Interest 41 (Fall).

PARENTI, M. (1970) "Power and pluralism: a view from the bottom." Journal of Politics 32 (3): 501-536.

PATEMAN, C. (1970) Participation and Democratic Theory. Cambridge, England: Cambridge University Press.

PITKIN, H. F. (1972) The Concept of Representation. Berkeley: University of California Press.

PREWITT, K. and H. EULAU (1969) "Political matrix and political representation: prolegomenon to a new departure from an old problem." American Political Science Review 63 (2): 427-441.

ROSSI, P. H., R. A. BERK, and B. K. EIDSON (1974) The Roots of Urban Discontent. New York: John Wiley.

SCHUMAKER, P. D. (1981) "Political responsiveness under different structural and cultural conditions: an examination of a deductive model." Presented at the Midwest Political Science Association, Cincinnati, April 17.

——— (1977) "Synthetic estimates of citizen policy priorities for American communities." Presented at the annual meeting of the Western Political Science Association, April 2.

——— and D. BILLEAUX (1978) "Group representation in local bureaucracies." Administration and Society 10 (4): 285-316.

SCHUMAKER, P. D. and R. W. GETTER (1977) "Responsiveness bias in 51 American communities." American Journal of Political Science 21 (2): 247-281.

SCHUMAKER, P. D. and B. LOOMIS (1979) "Responsiveness to citizen preferences and societal problems in American communities," pp. 38-66 in S. M. Hines and G. W. Hopkins (eds.) South Atlantic Urban Studies, Vol. 3. Columbia: University of South Carolina Press.

SCHUMAKER, P. D., R. W. GETTER, and T. N. CLARK (1979) Policy Responsiveness and Fiscal Strain in 51 American communities: A Manual for Studying City Politics Using the NORC Permanent Community Sample. Washington, DC: American Political Science Association.

SEIDMAN, D. (1975) "Simulation of public opinion: a caveat." Public Opinion Quarterly 39 (3): 331-342.

SHAFFER, W. R. and R. E. WEBER (1974) Political Responsiveness in the American States. Beverly Hills, CA: Sage.

SHARKANSKY, I. (1971) "Economic theories of public policy: resource-policy and need-policy linkage between income and welfare benefits." Midwest Journal of Political Science 15 (4): 722-740.

SUTTON, R. (1973) "The states and the people: measuring and accounting for 'state representativeness.' " Polity 5 (4): 451-476.

VERBA, S. and N. H. NIE (1972) Participation in America: Political Democracy and Social Equality. New York: Harper & Row.

WAHLKE, J. C. (1978) "Policy demands and system support: the role of the represented," pp. 73-90 in H. Eulau and J. Wahlke (eds.) The Politics of Representation. Beverly Hills, CA: Sage.

WEBER, R. (1975) "The political responsiveness of the American states and their local governments," pp. 189-225 in L. N. Reiselbach (ed.) People vs. Government: The Responsiveness of American Institutions. Bloomington: Indiana University Press.

——— (1971) Public Policy Preferences in the States. Bloomington: Institute for Public Administration, Indiana University.

——— and W. R. SHAFFER (1972) "Public opinion and American state policy-making." Midwest Journal of Political Science 16 (4): 683-699.

WEISSBERG, R. (1978) "Collective vs. dyadic representation in Congress." American Political Science Review 72 (2): 535-547.

WILSON, J. Q. and E. C. BANFIELD (1971) "Political ethos revisited." American Political Science Review 65 (4): 1048-1062.

WOLFINGER, R. E. (1971) "Non-decisions and the study of local politics." American Political Science Review 65 (4): 1063-1080.

ZEIGLER, L. H. and H. J. TUCKER (1978) The quest for Responsive Government. North Scituate, MA: Duxbury Press.

ZEIGLER, H., M. K. JENNINGS, and G. W. PEAK (1974) Governing American Schools: Political Interaction in Local School Districts. North Scituate, MA: Duxbury Press.

Part V

Population Redistribution and the Location of Economic Activities

11

Population and Economic Base Changes in Metropolitan Areas

JOHN D. KASARDA

DURING THE PAST two decades much has been written about the changing demographic and economic structure of metropolitan America and numerous empirical investigations have been conducted (see Frisbie, 1980; Goldstein, 1980, for literature reviews). These studies, unfortunately, have been of limited utility for explicating the structural roots of contemporary urban economic problems. Most have focused on specific population or economic base changes in relative isolation of one another and have restricted analysis to a case study variety or predetermined selection of metropolitan areas, usually the largest and oldest. There remains a pressing need for a comprehensive and up-to-date analysis of metropolitan structural change, with special emphasis on those factors exacerbating urban economic problems and fiscal strains. Such a study should not be a shotgun approach or fishing expedition, however. Problem formulation, data acquisition, and empirical analysis should be grounded in a theoretical framework that specifies both the interdependencies and disarticulations of population and economic base changes and elaborates their consequences for the economic vitality of our cities and their inhabitants.

In this chapter such a framework will be outlined. Guided by this framework, I will then suggest some important urban policy research questions and describe a variety of available secondary data bases which, when brought together, could help answer these questions.

THE EXPANSION FRAMEWORK

The framework to be outlined herein is derived from the theory of ecological expansion as originally formulated by McKenzie (1933) and

Hawley (1950) and elaborated in Kasarda (1972a, 1972b, 1976, 1978). In brief, this theory stipulates that improvements in short-distance transportation and communication in large urban areas foster a selective redistribution of population and economic functions, which increases both territorial differentiation of various resident groups and activity patterns and their functional interdependence over an ever-widening, more deconcentrated area. The underlying assumption is that as population and economic activities deconcentrate from center to periphery, contact is maintained with the center on a routine basis. This contact may be direct, as in terms of journey-to-work from suburban residence to central-city job, or indirect, such as suburbanites conducting business at branch banks or suburban retail outlets whose headquarters are located in the central city. If the metropolitan system is to be maintained as a functionally integrated unit, expansion of the periphery will typically be matched with a development of administrative, communication, professional, technical, and other specialized support functions in the center which coordinate and integrate activities and relationships throughout the expanded system. If the system is becoming more open and polycentric, this tight correspondence between peripheral expansion and core development may not emerge.

Consistent with conventional urban ecological theory, the expansion of the metropolis involves both centrifugal and centripetal movements that are selective by population group and economic activities. In the contemporary American metropolis, centrifugal movements involve the dispersion to the suburban and exurban areas of middle-income groups, industrial facilities, and commercial establishments providing consumer goods and services. Centripetal movements entail the concentration in the urban core of lower-income population, administrative and central office functions, communication and financial functions, and specialized business, professional, and technical services. The centrifugal and centripetal movements result in a diffuse metropolitan structure exhibiting two general loci: (1) an inner locus (or central city) which, while declining as a residential and commercial base, continues to function as the primary communication, administrative, financial, professional, and technical hub for the expanded system; (2) a multinucleated outer locus (or suburban ring) whose expanding resident population complements the central city by periodically drawing on its goods and services and by staffing many of the city's economic institutions, particularly their professional and middle- and upper-level management positions.

The essence of the expanded metropolitan system is its dispersed, loosely coupled network of residential, commercial, and industrial nodes whose linkages and interactions are sustained by advanced transportation and communication technologies. Its dynamics include the perpetual generation of new suburban and exurban housing developments (some as far as 100 miles from the metropolitan core); steady augmentation of peripheral industrial parks and shopping malls; increased outer zonal traffic flows among the deconcentrated residential, industrial, and commercial nodes; greater amounts of suburban ring to central-city commuting of white-collar workers; increased reverse commuting (city residence to suburban ring job) especially by blue-collar workers as their employment opportunities deconcentrate and as computer-age cities structurally transform from centers of material goods production to centers of information exchange and service consumption (Kasarda, 1980, 1981).

EXPANSION, INNER-CITY UNEMPLOYMENT, AND FISCAL STRAIN

How is this expansion framework and the selective redistribution of population and economic activities it entails specifically related to current urban economic problems? While these problems must assuredly have multifaceted and complex determinants, their structural roots lie in the centrifugal and centripetal movements inherent in the expansion process. These movements, which are analytically separable but interrelated, include (1) the urban exodus of middle- and upper-income families to the suburbs and exurbs; (2) the influx into the central cities of poor minority groups, chronically unemployed, the aged, social misfits, and others who tend to be more of a fiscal liability than an asset to city budgets; (3) the redistribution of the bulk of metropolitan commerce and industry from central cities to suburban rings, which has eroded city tax bases, drained the cities' retail, wholesale, and manufacturing employment sectors while making blue-collar jobs inaccessible to many inner-city residents; (4) the buildup in the central cities of administrative office complexes, financial institutions, specialized professional and technical services, and other "knowledge class" industries whose jobs require substantial education and training and, hence, are simply inappropriate to the growing concentrations of less educated inner-city residents; and (5) the daily flow into the central city of large numbers of suburbanites who make regular use of central-city public services and facilities as part of their journey-to-work and shopping and recreational activities.

As a consequence of these interrelated movements, central cities are facing greater public service demands, while their own ability to pay for additional public services is deteriorating. This widening resource-service demand gap has given rise to growing geopolitical welfare, as federal and state governments increasingly transfer tax revenues on an areal basis from more economically viable suburban and exurban communities to less viable central cities. Indeed, losses of the middle-income population and commercial and industrial facilities from older, larger central cities have been so substantial that only through growing infusions of intergovernmental transfer payments and subsidies (such as revenue sharing, block grants, and federal jobs programs) have these cities been able to maintain basic public services and keep their disadvantaged residents economically afloat. The fundamental problem, then, is not one of insufficient metropolitan resources, but that, in the process of metropolitan expansion, the majority of the metropolis' demographic and economic resources has shifted from the central cities to the politically autonomous suburban rings, and beyond.

ARE BASIC EXPANSION TRENDS REVERSING?

Recently, a number of social scientists and urban journalists have claimed that some of the traditional centrifugal and centripetal movements of metropolitan expansion are reversing. Basing this claim primarily on personal observation and case studies of small, urban subareas, they argue that white, middle-class residents are streaming back to inner-city neighborhoods and that our cities are showing remarkable new signs of residential and commercial health. Indeed, a few have gone so far as to suggest that we are in the midst of an urban residential and commercial renaissance.

Without doubt, selective residential and commercial revitalization is occurring in virtually all major cities. This in itself is an encouraging sign. However, carefully conducted surveys and research to date show that revitalization currently is affecting relatively small portions of cities and that the centrifugal and centripetal movements specified in expansion theory still predominate. The main findings of these surveys and analyses document the following:

(1) The overwhelming flow of middle-class households during the 1970s has been out of, not into, the central cities. For every household with income levels of $15,000 or more in entering central cities between March 1975 and March 1978, three left the cities (U.S. Bureau of the Census, 1978). During this period, four out of five family heads who left the central cities had incomes over $15,000 per annum. As a result, aggregate personal income of central-city residents continues to drop precipitously. The aggregate income difference between movers out of the central city and movers into the central cities between 1970 and 1975 resulted in a loss of $29.6 billion in personal income for the cities. The aggregate income

differences between movers out of and movers into the central cities between 1975 and 1978 led to another $18 million city loss. Sterlieb and Hughes (1980) computed the cumulative effect of central-city migration between 1970 and 1978 on the aggregate personal incomes of central-city residents. They found that during this period the central cities experienced a net loss of $64.8 billion (in constant 1976 dollars) due to the differential income levels between immigrants and outmigrants and the greater number of outmigrants than immigrants. Analyzing changes in the central-city suburban income gap during the 1970s, Long and Dahmann (1980) show that the gap widened in terms of both family income levels and per capita income.

(2) Revitalization has occurred in only a tiny fraction of any city's neighborhoods and has affected only a small portion of its housing stock. Between 1968 and 1979, only one-half of one percent of the nearly 20 million housing units in cities were affected by revitalization or restoration efforts (U.S. Department of Housing and Urban Development, 1979).

(3) The vast majority (more than 70 percent) of households residing in revitalized central-city units were intracity movers and not suburbanites or nonmetropolitan residents who moved back to the city (James, 1980).

(4) The concentration of minorities and poverty-level population in central cities is continuing to increase, in both absolute and proportional terms. Between 1970 and 1978 central cities lost nearly four million white residents, whereas the black population in the central cities increased by 542,000. Despite a net national decline of 2.2 million persons below the poverty level between 1970 and 1978, the total number of central city residents below the poverty level continued its dramatic growth. Included in this growth were a net increase of 441,000 blacks below the poverty level and a net increase of 370,000 female-headed households below the poverty level (U.S. Bureau of the Census, 1978).

(5) Substantial residential and commercial disinvestment continues to be active in almost all older, larger central cities and far exceeds reinvestment. During the 1970s, the expansion of blighted neighborhoods and the increase in residential abandonment that characterized these cities were matched with accelerated deterioration and abandonment of commercial and industrial facilities (Sternlieb and Hughes, 1980).

(6) The growth of office employment in the central business districts during the 1970s did not nearly compensate for central-city losses of blue-collar jobs and employment declines in the retail and wholesale sectors. As a result, the overall employment base of most large cities is continuing to deteriorate. In New York City, for example, although white-collar professional, clerical, and managerial jobs were expanding in Manhattan during the 1970s, New York City as a whole lost more than 600,000 jobs during the decade—the vast bulk in the blue-collar sector. This net job loss alone resulted in a tax loss to New York City of almost $500 million (Embry, 1979), an amount that would have more than covered its 1980 budget deficit.

In sum, many major U.S. cities are still in deep trouble as their overall middle-income residential bases and job opportunities continue to shrink. This is especially the case for the larger, older cities in the northeast and northcentral regions of the country. However, the same signs of economic distress are also beginning to appear in many younger, smaller metropolitan central cities and those in the South and the West.

NEEDED RESEARCH AND DATA SOURCES

As stated in my opening paragraph, we badly need a comprehensive and up-to-date empirical analysis of demographic and economic base changes that have occurred in *all* our metropolitan areas since World War II. This analysis would document in detail the full range of redistributional movements I have noted and assess their impact on such factors as employment and housing opportunities in the central cities and suburban rings, commuting and reverse commuting, urban fiscal resources, public service expenditures, and municipal debt. More specifically, the following questions, and others related to them, could be answered:

(1) What effect has the selective spatial redistribution of economic opportunities had on those who reside in the central cities, particularly poor minority groups and blue-collar ethnics with substandard educational backgrounds? One's working assumption might be that the centrifugal and centripetal forces of expansion have exacerbated problems of central-city unemployment and poverty by redistributing blue-collar employment opportunities to the suburbs and replacing them, in part, with white-collar jobs that require educational levels and professional skills beyond those of large numbers of inner-city residents.

(2) What effect have the centrifugal and centripetal movements of population and economic activity had on the fiscal resources of central cities and suburban rings? Here, one could assess the degree to which deconcentration of middle-income population and economic activities since World War II have eroded central-city tax bases and have created a resource imbalance between central cities and their suburban rings. In addition, one might examine the differential welfare burdens assumed by central cities as a consequence of the growing concentration of minorities, low-income families, and unemployed in the metropolitan cores.

(3) To what extent have changes in metropolitan structure since World War II been responsible for increased public service expenditures incurred by central cities? Previous research has shown that, as middle- and upper-income groups have shifted their residences to the suburbs, many continue to routinely use central-city services and facilities as part of their journey-to-work and shopping and recreational activities. One could investigate

the marginal costs borne by central-city governments as a result of public services they provide to their suburban neighbors *and* the additional revenue cities receive from suburbanites who use city facilities. The objective here would be to shed empirical light on the policy-packed central-city suburban exploitation issue.

To conduct the suggested research, it would be necessary to bring together data from a variety of secondary sources. Basic demographic changes occurring within central cities and suburban rings (for example, population growth, changing density patterns, and changing racial, occupation, and income composition) could be documented utilizing the 1940, 1950, 1960, 1970, and 1980 *Censuses of Population* and numerous *Current Population Surveys.*

The changing distribution of retail, wholesale, and business service activities (including number of establishments, employees, and sales or receipts, by nature of business) could be analyzed with data from the 1939, 1948, 1954, 1958, 1963, 1967, 1972, 1977, and 1982 *Censuses of Business,* supplemented by information provided in *County Business Patterns,* 1946 to the present. The suburbanization of manufacturing activity (including number of establishments, production and other manufacturing employees, and value added) could be assessed with data obtained from the 1939, 1947, 1954, 1963, 1967, 1972, 1977, and 1982 *Censuses of Manufacturers* and the *Annual Survey of Manufacturers,* 1958 to the present.

The centripetal buildup in the central cities of governmental, mangerial, professional, clerical, and other white-collar functions could be documented with published and unpublished data from the place-of-work reports of the 1960, 1970, and 1980 *Censuses of Population.* These data could be supplemented with data on the growth of administrative and central office functions in central cities and suburban rings that could be provided by such organizations or associations as the Building Owners and Managers Association, International (Chicago, Illinois).

The nature and extent of commuting between central cities and suburban rings (including reverse commuting from city to suburb) by place of residence, race, income, occupation, and industry of commuters could be examined with published and unpublished data on journey-to-work derived from the 1960, 1970, and 1980 *Censuses of Population.* More detailed social, occupational, and demographic data on commuters by place of residence and place of work could be obtained from the 1960, 1970, and 1980 compatible 1/100 Public Use Samples.

Relative changes in central-city and suburban revenue by source (for example, property taxes, sales taxes, intergovernmental grants) and public service expenditures by function (such as welfare payments and expenditures for police, fire, highway, sanitation, recreation, administration, and so on) could be examined with public finance data obtained from the 1957, 1962, 1967, 1972, 1977, and 1982 *Censuses of Governments,* the annual compendia of *City Government Finances,* 1947 to 1981, and the *Municipal Year Book,* 1945 to 1981. These data could be supplemented by specific information on revenue and public services, available in published reports of local governments in metropolitan areas.

Data on employment status by race, education, occupation, and industry of employment (or last job held by unemployed) would be obtained from publications of the 1950, 1960, 1970, and 1980 *Censuses of Pupulation* and from periodic reports of the Bureau of Labor Statistics. The 1970 and 1980 *Census of Population* fourth-count summary tapes would be particularly useful in locating high unemployment sectors and poverty pockets within each metropolitan area and examining their ecological characteristics. In addition, the 1960, 1970, and 1980 Public Use Samples provide data for individuals' employment status by place of residence, place of work (central-city/suburban ring dichotomy), occupation, industry of employment, and a broad variety of other useful variables such as race, income, and educational background.

The metropolitan communities for which the majority of data should be gathered and analyzed are all Standard Metropolitan Statistical Areas currently delimited by the Bureau of the Census (N = 270). To improve longitudinal comparability, individual SMSA boundaries should be adjusted to constant areas over time according to the latest delimitations. Since SMSAs are comprised of counties (except in New England, where towns and cities are used), adjustments could be made by adding or deleting appropriate county data at a particular date. The ecological structure of SMSAs created since 1950 could be reconstructed historically on the basis of their county and place data for prior decades. One should reconstruct SMSA boundaries for New England in terms of counties to improve comparability with SMSAs in other regions of the nation.

It would also be necessary to adjust the census data for central cities and suburban rings to account for intrametropolitan boundary changes over time. For example, annexation and consolidation systematically understate suburban ring growth and overstate central-city growth. Methodological procedures for taking into account annexation are presented in Kasarda (1976). When demographic and economic data do not correspond temporally (for example, manufacturing data for 1963 or 1972 and population data for 1960 and 1970), standard interpolation and extrapolation could be carried out to adjust the data to comparable dates. Moreover, in all longitudinal analyses of economic activity utilizing monetary indicators (sales, receipts, value added, and public service expenditures, for example), the indicators should be converted to constant dollars using appropriate price deflators.

Indexes based on Lorenze curves and location quotients could be used to measure the centrifugal and centripetal movements of the different population groups and economic activities within SMSAs. In some instances, the nature of the published data will require that the indexes be constructed on the basis of central-city and suburban ring differences. However, where more geographic detail is provided, concentric-type distance zones and smaller territorial aggregations could be used to assess redistributional patterns and processes.

One should examine the centrifugal and centripetal movements of people and economic activity for the total population of SMSAs and within subcategories of SMSAs. The specific categories should be determined on the basis of theoretically relevant criteria. In the past, such factors as SMSA age, size, density, region, and availability of mass transportation have been used. The existing research literature would lead one to suspect that significant variation in ecological change will be observed within the different types of SMSAs.

Much more could be said here regarding potential data bases and analytical methods. Suffice it to say that the wealth of longitudinal data available from the secondary sources discussed above, together with current computer technology, provide us with a unique opportunity to assess the economic and fiscal implications of metropolitan expansion and structural change for our central cities and suburbs.

REFERENCES

EMBRY, R. C., Jr. (1979) Statement prepared for the Subcommittee on Fiscal and Intergovernmental Policy of the Joint Economic Committee of the Congress of the United States, March 20. Washington, DC: Government Printing Office.

FRISBIE, W. P. (1980) "Urban sociology in the United States: the past 20 years." American Behavioral Scientist 24 (November/December): 177-214.

GOLDSTEIN, G. S. (1980) "Recent developments in urban economics." American Behavioral Scientist 24 (November/December): 228-276.

HAWLEY, A. H. (1950) Human Ecology: A theory of Community Structure. New York: Ronald Press.

JAMES, F. J. (1980) "The revitalization of older urban housing and neighborhoods," pp. 130-160 in A. P. Solomon (ed.) The Perspective City. Cambridge: MIT Press.

KASARDA, J. D. (1981) "To revitalize the cities: in search of an urban policy." American Demographics 3 (January): 13-20, 44-48.

——— (1980) "The implications of contemporary redestribution trends for national urban policy." Social Science Quarterly 61 (December): 373-400.

——— (1978) "Urbanization, community, and the metropolitan problem," in D. Street et al. (eds.) Handbook of Contemporary Urban Life. San Francisco: Jossey-Bass.

——— (1976) "The changing occupational structure of the American metropolis: apropos the urban problem," pp. 113-135 in B. Schwartz (ed.) The Changing Face of the Suburbs. Chicago: University of Chicago Press.

——— (1972a) "The impact of suburban population growth on central city service functions." American Journal of Sociology 77 (6): 1111-1124.

——— (1972b) "The theory of ecological expansion: an empirical test." Social Forces 51: 165-175.

LONG, L. H. and D. C. DAHMANN (1980) The City-Suburb Income Gap. Washington, DC: U.S. Bureau of the Census.

McKENZIE, R. (1933) The Metropolitan Community. New York: McGraw-Hill.

STERNLIEB, G. and J. W. HUGHES (1980) America's Housing : Prospects and Problems. New Brunswick, N.J: Center for Urban Policy Research, Rutgers University.

U.S. Bureau of the Census (1978) "Geographical mobility: March 1975 to March 1978," Current Population Reports Series P-20, No. 331. Washington, DC: Government Printing Office.

U.S. Department of Housing and Urban Development (1979) "Whither or whether urban distress." Working paper presented before the Joint Economic Committee, Congress of the United States, March 20. Washington, DC: Government Printing Office.

12

Putting Politics into Urban Ecology: Estimating Net Effects of Zoning

ANNE B. SHLAY
PETER H. ROSSI

URBAN ECOLOGY seeks to portray a dynamic picture of urban social change. Asking how people organize and adapt themselves to the physical constraints within an urban setting, the ecologist examines the mechanisms by which the spatial organization of community is created and maintained within the metropolis (Hawley, 1950). Specifically, an ecological focus looks at the relationships among positions in the urban space (locations), positions in the urban social structure (class, race, occupation, education, and so on) and the distribution of land use (Park, 1967; Schwirian, 1974a; Berry and Kasarda, 1977).[1] Hence, the mechanisms which "sort" people into so-called natural areas explain the way a social structure is reproduced within urban society.

The social structure to which urban ecology traditionally has been attentive is strangely apolitical. To be fair, it should be noted that in the 1920s, when ecological frameworks originated, local governments made few, if any, direct attempts to regulate the uses of land. Transportation franchises were given to entrepreneurs and street patterns were laid out, but today's complex of land use regulations and building codes were in only embryonic stages. Unregulated land use characterized the Chicago over which Park and Burgess superimposed concentric zones centered on

AUTHORS' NOTE: *This chapter was prepared under Grant No. HUD 2367 from the Department of Housing and Urban Development. We are indebted to Jeffrey K. Liker, Lisa Sushelsky, Eleanor Weber-Burdin, and James D. Wright for their devoted help at crucial points.*

ne Loop, and today's ecologists persist in slighting the importance of governmental regulations in urban development and the maintenance of land use patterns (Schwirian, 1974b; Berry and Kasarda, 1977).[2]

The purpose of this chapter is to bring local politics into the understanding of land use distributions. The nonpolitics of unregulated land use cleared the path for early ecologists to discuss "natural" distributions of populations in terms of competition, invasion, succession, and the like, because these were the factors that clearly operated to determine the spatial arrangements during a period of unregulated industrial and urban development. As politics were separate from industrial development, politics were also distinct from the social processes which formed our cities. Therefore, the traditional ecological models posed empirically accurate descriptions of urbanization in this earlier era. These descriptions, however, are likely to be dated within an era or stage of urbanization when the politics which regulate land use can possibly take precedence over the "natural" unfettered social processes of competition and succession.

The relative absence of politically distributed areally based goods and services during ecology's formative stages also influence other aspects of ecological theory. For the traditional ecologist, the value of a particular location was based on its accessibility to the central business district (Hawley, 1950), not on characteristics associated with the neighborhood per se. Hence, ecological models did not address the consequences of differential locational patterns for varying social groups. Without denying the importance of accessibility to the central business district, we suggest that urban location provides differential access to public and private goods and services. That is, urban locations have varying degrees of access to employment, education, consumer goods, crime, housing, police and fire protection, snow removal, recreational facilities, street lighting, and so on. Thus urban locations can also be viewed as indicators of class and, more important, as indicators of life chances. Residential mobility can be viewed in part as social mobility, a movement which reduces or enhances one's class or status position, one's life chances, in essence—a movement which changes one's position in what may be termed an "urban opportunity structure."[3]

Conceiving the collectively consumed benefits accrued through location in this way suggests a key relationship between location and mobility. First, differential access to areally based benefits is part of the urban stratification process. Second, residential location is linked to differential achievement capacities and therefore influences social mobility. Finally, urban social mobility is politicized to the extent that areally based benefits are publicly appropriated and politically distributed. Therefore, the social-spatial distribution denotes relative positions within the urban opportunity structure which heightens the importance of those mechanisms creating congruent social and spatial patterns.

To acquire a relevant contemporary analysis of the relationships among urban political, social, and spatial dimensions, this chapter extends traditional ecology by including among the mechanisms which reproduce the urban social structure the relationship of a city's land use to the politics that govern land use. This study takes one major part of the local political process, zoning, and looks at its impact on metropolitan development by viewing zoning as an important determinant of land use.

ZONING AND EXCLUSION

Zoning as a means to ensure "a place for everything and everything in its place" is an accepted practice within most local governments (Perin, 1977).[4] A process which designates land parcels by law for specifically designated uses, zoning can segregate single-family homes from apartments, separate particular areas for industry, and create special areas for shopping or commercial purposes. Zoning also aften regulates the size of structures and building lots. Termed "bulk and area" requirements, these regulations typically stipulate the maximum building height and minimum dimensions for the building lot area, lot frontage (width), floor area, and front, side and rear yards. Such requirements are often specified for each type of land use to the point of regulating the minimum amount of land required per extra bedroom.

Over a half-century ago, zoning was proposed as a method to rid New York City's Fifth Avenue luxury retail establishments of adjacent garment factory lofts and, at the same time, was put forward as a means of reforming the slums (Friedman, 1968; Toll, 1969; Fitch, 1977). Zoning advocates displayed goals that ranged from a desire mainly to improve housing conditions for the poor to overt "classism." Liberal reformers and upper-class conservatives could find common cause in zoning. All were easily subsumed under the broader issue of "land use incompatibility." Convinced that the mixing of factories with residences was detrimental to families' "health, safety, and welfare," in 1926 the Supreme Court ruled that zoning was constitutional under the broad mandate of police power (Euclid v. Ambler Realty Co., 1926).[5] In essence, local power to control land use for the good of family life appeared more important than the parochial interests of individual property owners.

Because defining zoning as a broad welfare measure without planning direction would lead to arbitrary and eclectic local definitions of land use imcompatibility, the application of zoning was circumscribed by the court by a clearly articulated scheme of land use priorities, the "Euclidean Pyramid."

Named after the Supreme Court decision in which this scheme was first developed, Euclidean zoning designated a rank ordering of land use desirability to guide local governments in their decisions whether to segregate or to mix particular types of land use.[6] According to the scheme, zoning's primary purpose is to protect single-family dwelling neighborhoods, placed at the top of the priority pyramid. Single-family land use is followed by more "permissive" residential land uses, such as duplexes and apartment houses. The increasing width of the pyramid indicates a successive accumulation of land uses with greater permissiveness. Thus the base of the pyramid is defined by industrial zoning, which permits all types of land use. Valuing single-family dwellings as the highest and best use" became the common local method for prioritizing land use.

Euclidean zoning's definition of "good neighborhoods" (single-family housing) versus "bad neighborhoods" (mixed land use) is strengthened by a number of additional rationales. First, as a "protectionist device," zoning is justified as a public welfare measure. Zoning's benign intent is to protect residential neighborhoods from the congestion, noise, traffic, pollution, and general ugliness associated with commerce and industry. Zoning seeks to *protect the rights of all individuals* by segregating land use in a manner that maximizes the well-being of the entire community (Friedman, 1968; Williams, 1975).

Second, zoning is designed to "protect" property values on behalf of consumers. By segregating perceived deleterious land use, zoning acts as a brake on market forces. As an economic policy, zoning distributes land use in a manner so that *property will have maximum value on the market* (Babcock, 1969).[7]

Third, zoning works to "internalize externalities"—that is, to segregate "lower" types of land use (industry, commerce, and apartments) from the "highest" type of land use (single-family homes.)[8] In this way, localized externality effects are negated because they are circumscribed within specifically designated areas, and those land users who generate externalities are thereby constrained to bear them.

Finally, as a method to protect the public's welfare and, at the same time, to protect property values, zoning becomes a local planning policy. As a "tool of planning" (Haar, 1959; Williams, 1966; Dalafons, 1969; Babcock, 1969), zoning is juridically and practically justified as a method to be used "in accordance with a comprehensive plan." By combining land use regulations with comprehensive planning, zoning is intended to *guide land use in accordance with perceived long-range local needs.*

With experience, these principles have also become tied to the less abstract goal of managing local tax revenues (Williams and Norman, 1971). Local governments must provide many goods and services—for example, schools, roads, sewers, street lights, water, parks, snow removal, and police protection—whose primary funding sources are local property taxes. Through zoning, local government can limit per capita expenditures and also can expand local tax bases. To preserve the taxable value of land while limiting the tax rate, a community may exclude potential negative externalities and simultaneously fence out land users who might not "pay their way" (Babcock, 1969; Williams and Norman, 1971; Danielson, 1976; Moskowitz, 1977).

Termed "fiscal zoning," this type of policy directly affects residential development. Used as a fiscal policy, zoning can prohibit or severely restrict high-density developments because such uses often cause net increases in government spending. Apartments occupy less land than single-family dwellings, so that the tax returns per apartment unit may be lower than for tax per unit of detached housing. Apartment residents, however, may require levels of service delivery similar to low-density land users, particularly if apartments house families with school-age children. Therefore, local communities often believe that higher density development causes per capita spending to increase relative to revenues (Babcock, 1969; Williams and Norman, 1971; Bergman, 1974; Danielson, 1976; Moskowitz, 1977).

Of course, fiscal zoning affects more than the mere flow of money in and out of City Hall. To achieve fiscal goals, a community ipso facto regulates its socioeconomic composition. In this way, fiscal zoning is also a social planning policy that may be directed at or result in socioeconomic exclusion. Termed "exclusionary zoning," such practices indirectly operate as methods for *legal* socioeconomic and racial discrimination. Therefore, zoning to maximize a community's public welfare simultaneously works to "maintain a community's residential status quo.[9] In other words, zoning is a method for "keeping up the neighborhood."

Suburban communities often have been charged with following exclusionary zoning policies under the rationale of fiscal or "no growth" goals (Downs, 1970; Babcock and Bosselman, 1973; Bergman, 1974; Danielson, 1976; Moskowitz, 1977; Perin, 1977). These practices are claimed to create an "invisible wall" which restricts the shifting of low-income families from older, central-city neighborhoods to newly developed suburban communities. Hence, local land use is viewed as maintaining socioeconomic and racial homogeneity of suburban areas and concentrating low-income households and racial minorities within already fiscally troubled central cities. Zoning has influence that goes beyond affecting development patterns of single neighborhoods or communities; it can affect the social-spatial structure of entire metropolitan regions. Thus zoning may work to segregate the suburban "rich" from the urban "poor," and whites from blacks and other minorities.

DESIGN AND DATA BASE

This chapter looks at how zoning regulates the urban social-spatial structure by examining the impact of zoning on housing and population growth over time. Using longitudinal data, these analyses model the net effects of zoning within one major metropolis—the Chicago SMSA over the decade 1960-1970. To examine the impact of zoning on social-spatial processes, we use a longitudinal design, one that links zoning classifications and changes in the housing and population stocks over the period 1960 to 1970 for a sample of census tracts drawn from the Chicago SMSA.[10]

For the purposes of this study, census tracts are used as proxies for neighborhoods. Since tract boundaries are drawn with attention to population size and socioeconomic composition, they are approximately equal in population and are intended to be socioeconomically homogeneous. Using census tracts as the units of analysis, we assess the impact of zoning on neighborhood changes over the decade while holding constant the initial population and housing stocks in the tracts studied.

This study relies on two data sources: the 1960 zoning regulations of the city of Chicago, selected suburbs, cities, and counties within the Chicago SMSA; and census tract data from the 1960 and 1970 censuses.

Our sample of census tracts was drawn in two stages.[11] First, the Chicago SMSA was divided into two parts: the city of Chicago and the rest of the SMSA. Second, central-city tracts were stratified by two characteristics: housing density and mean household income levels. Tracts were sampled within strata in order to maximize variance in zoning.[12] In selecting suburban tracts, we sampled political subdivisions in order to maximize the zoning practice variation among political units. Suburban municipalities were stratified according to the total number of tracts within each political unit—municipalities containing one tract, two to four tracts, five to nine tracts, and ten or more tracts. Municipalities were randomly selected from each stratum so that the suburban sample consisted of the distribution of tracts contained within the political units.

The zoning regulations for sampled tracts were obtained by site visits to the zoning agencies for the political subdivisions involved. Eliminating tracts for which no census data and/or zoning data could be obtained left a final sample of 395 census tracts: 249 within the central city and 146 tracts shared among 39 local political subdivisions and 2 counties (See Appendix).[13]

MEASURING ZONING

Land use regulation in Illinois is left to local communities to define. As a consequence, local zoning regulations are particularistic and highly variable. Specific regulations vary from place to place, although, fortunately, overall Euclidean pyramid principles prevail.

To make local *residential zoning districts* comparable across communities, land use categories were grouped by the most common rules used: permitted land uses within each district and minimum lot areas per dwelling unit. The resulting eleven standardized residential zoning categories are ordered, as expected, according to the Euclidean prioritization scheme in which single-family dwellings are the "highest and best use" and low densities are "higher" than high-density regulations.

The use and area requirements defining each standard zoning level are shown in Table 12.1. Note that all single-family dwelling zoning categories permit only single-family dwellings, all two-family dwelling zoning categories permit both duplexes and single-family dwellings, and all multiple-family dwelling zoning categories permit all types of residential land use. Hence, these levels range from the most restrictive zoning category, which permits only single-family dwellings on large lots, to the most permissive zoning category, which permits any type of residential land use on very small lots.[14]

To use these residential zoning levels, the zoning classifications used in each local community were examined in detail and each residential category was placed in the "standard" coding category closest to the local definition. No difficulty was found in coding residential zoning: Of the 42 communities studied, all of the residential zoning categories could be fitted into the standard scheme shown in Table 12.1.

The Euclidean scheme also made it possible to make *nonresidential* zoning districts comparable. Standardized nonresidential zoning categories were created to reflect differential potential negative externalities.[15]

TABLE 12.1 Standard Residential Zoning Levels Code

Single-Family Dwelling Zoning Categories

Symbol	Minimum Lot Area*	Weight Used
S-1	43,561 to 217,800 sq. ft.	11
S-2	21,800 to 43,560 sq. ft.	10
S-3	10,001 to 21,799 sq. ft.	9
S-4	7,201 to 10,000 sq. ft.	8
S-5	0 to 7,200 sq. ft.**	7

Two-Family Dwelling Zoning Categories

Symbol	Minimum Lot Area*	Weight Used
T-1	3,000 to 6,500 sq. ft.	6
T-2	0 to 2,999 sq. ft.**	5

Multiple-Family Dwelling Zoning Categories

Symbol	Minimum Lot Area*	Weight Used
M-1	2,001 to 4,000 sq. ft.	4
M-2	1,001 to 2,000 sq. ft.	3
M-3	501 to 1,000 sq. ft.	2
M-4	0 to 500 sq. ft.**	1

*Minimum lot area per dwelling unit for the primary permitted land use within each category, e.g., minimum lot area per dwelling unit within a multifamily structure.
**"0" does not mean that no lot is required, but rather than no minimum lot area is specified.

The overall residential zoning status of each tract was measured by mapping tract boundaries on zoning maps. The percentage of area zoned for residential use in each tract was calculated for each of the standard zoning levels shown in Table 12.1.[16] Using the weights shown in that table, an "average" residential zoning level was computed for each tract and defined as the Exclusionary Zoning Index.[17] The Exclusionary Zoning Index ranges from 0 to 11, a low score representing tracts containing largely permissively zoned residential land and a high score representing tracts containing quite restrictively zoned residential land.

Three zoning measures were computed to account for the zoning of entire tracted areas: an Exclusionary Zoning Index, the percentage of a tract zoned for business land use, and the percentage of a tract zoned for manufacturing land use. The distribution of zoning among central city and suburban tracts in 1960 is shown in Figure 12.1.

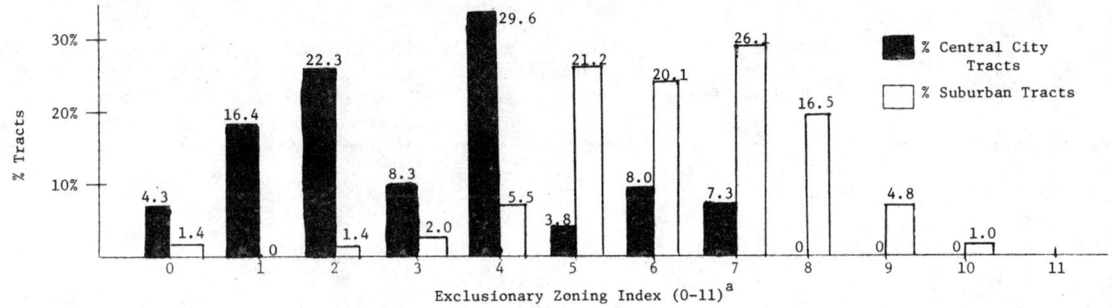

	Exclusionary Zoning Index			Percent Business[c]			Percent Manufacturing[d]		
	$\overline{X}$	SD	N	$\overline{X}$	SD	N	$\overline{X}$	SD	N
Central City[b] Tracts	3.36	1.79	363	23.2	17.0	356	31.1	25.3	186
Suburban Tracts	6.18	1.62	146	12.8	14.1	139	16.0	20.5	78

Figure 12.1 Distribution of Zoning within Central City and Suburban Tracts 1960

a. The index was computed to range from 0 to 11, where 11 is the most exclusive tract conceivable. Since the index is actually a continuous variable, the scores for each tract were rounded to the nearest whole number.

b. These figures are based on a weighted sample of central-city tracts. Low-income/high-density tracts and high-income/low-density tracts were undersampled in the city. By weighting these tracts by a factor of 1.78, these statistics describe a representative sample of central-city tracted areas. The weighting procedure produces an artificially inflated sample size. The already representative suburban sample of tracts required no weighting.

c. Means and standard deviations computed on the basis of only those tracts with some portion zoned for business land use.

d. Means and standard deviations computed on the basis of only those tracts with some portion zoned for manufacturing land use.

TABLE 12.2 Correspondence Between Zoning and Housing for Central-City and Suburban Tracts, 1960[a]

Independent Variables	(Dependent Variable is Number of Housing Units/Mile^2 1960)		(Dependent Variable is % Single-Family Dwellings 1960)		(Dependent Variable is % Multifamily Dwellings 1960)[b]	
	City	Suburb	City	Suburb	City	Suburb
Exclusionary Zoning Index[c]	–2,020.04*	–1,022.96*	11.63*	12.27*	–13.34*	–6.64*
	(250.50)	(95.88)	(.64)	(1.17)	(.64)	(1.10)
% Zoned for Business	42.60	–14.97	–.17*	–.29*	.28*	.23*
	(26.34)	(9.94)	(.07)	(.12)	(.07)	(.11)
% Zoned for Manufacturing	–88.76*	–25.87*	–.04	–.10	–.007	–.03
	(16.69)	(7.15)	(.04)	(.09)	(.04)	(.08)
Constant	16,120.39*	9,502.02*	–9.76*	–4.22*	91.71*	58.14*
	(1,255.47)	(684.23)	(3.20)	(8.34)	(3.22)	(7.89)
R^2 =	.39	.51	.63	.57	.71	.33
N =	(237)	(144)	(237)	(144)	(237)	(144)

*Significant at .05 level.

a. Entries are unstandardized regression coefficients with associated standard errors in parentheses.

b. Number of units within structures containing three or more dwelling units.

c. Index ranges from 1 to 11, where 1 is permissive residential zoning and 11 is exclusive residential zoning.

Compared to suburban tracts, central-city zoning was relatively permissive, with a strong majority of Chicago tracts (81 percent) receiving an index score of 4 or less. Suburban zoning was restrictive: 89 percent of the suburban tracts received a score greater than 4.[18] Nonresidentially zoned land was also more available within the central city. The average city tract zoned for either business or manufacturing land use allocated twice as much land for these purposes than did the suburbs.

These unsurprising city-suburb differences in the application of zoning provide one indicator that zoning has been effectively measured at the tract level. A second indicator is whether tract land use regulations correspond to the actual land use contained within these areas. Assuming that zoning regulations are obeyed and inforced, the way land is zoned should be directly related to local housing density and composition.

Table 12.2 shows the regression of 1960 tract housing composition (number of housing units per square mile, percent single-family dwellings, and percent multiple-family dwellings) on the three 1960 zoning variables. These "static" analyses do not represent causal processes but simply indicate the association between local zoning and land use. More exclusive residentially zoned tracts were less dense (more than 2000 fewer dwelling units per square mile in Chicago and more than 1000 units in suburban areas for each additional unit of the index) and contained proportionately more single-family homes than did permissive residentially zoned tracts within both the central city and the suburbs. Although housing density did not tend to vary significantly according to the amount of land allocated for business use, tracts zoned heavily for business contained fewer dwelling units (90 in Chicago and 26 in the suburbs) but more apartment houses (92 in Chicago and 58 in the suburbs) for each percentage increase in land zoned for manufacturing. The R^2 for each regression is high and significant, indicating that *the derived zoning measures effectively capture local zoning practices at the census tract level.*

MODEL SPECIFICATION

The analyses which follow estimate the effects of tract zoning in 1960 on changes in tract housing and population composition over the decade. The general model employed is as follows:

$$Y_{1970_i} = Y_{1960_i} + \Sigma b_i T_i + \Sigma b_j Z_i + e$$

where

Y_i is some 1960 or 1970 measure for tract i (e.g., number of housing units, median family income, number non-whites)

$\Sigma b_i T_i$ is a vector of tract i characteristics that relate to tract change

$\Sigma b_j Z_i$ is a vector of j zoning variables in 1960

e is the usual stochastic error term.

The purpose of the model is to capture the effects of the zoning regulations in force in 1960 on tract changes between 1960 and 1970. Since tracts tend to persist in character over time, do not change location, and have an existing physical plant, infrastructure, and the like, the 1960 characteristics of a tract are a very good predictor of their 1970 characteristics. Indeed, the 1960 characteristics embody the effects of the tracts' histories—namely, the effects over time of the many urban social processes that determine the tract characteristics at a point in time. The model therefore states that the effects of zoning are to be measured by deviations from expected growth patterns based on the persistence of charateristics over time (Bohrnstedt, 1969; Wright et al., 1979); Pendleton et al; 1980).[19] The coefficents for zoning measures derived from the model are estimates of the effects of zoning regulations net of the expected changes in the characteristics in question.

FINDINGS AND DISCUSSION

The "prosperous" 1960s was a period of intense housing development and also partly housing "undevelopment." Urban renewal provided central cities with the legal mechanisms and federal funding for local reconstruction involving the demolition of large numbers of housing units. Rapid development of vacant suburban land was fostered by low interest rates and improvements in intra-SMSA Highway systems.

Newly built suburban housing and an increase in family real income meant that for many urbanites, living in the city was no longer necessary or desirable. Newly prosperous or more prosperous urbanites took advantage of new suburban housing developments, leaving behind older, less valuable housing which was situated in a seemingly unpleasant, dangerous, and dirty environment. In addition, newly formed urban households found that housing "bargains" were located mainly in the suburbs.

At the same time, another population movement was taking place: the migration of southern blacks into northern industrial urban centers. Although the number of blacks moving into the city was large, these new urbanites barely replaced the large number of middle-class families moving out to the suburbs. The city gained a larger proportion of blacks but grew smaller in population. Table 12.3 summarizes these changes for the central city and suburban periphery from 1960 to 1970.

TABLE 12.3 Changes in Selected Housing and Population Characteristics for Central-City and Suburban Tracts, 1960-1970

	CITY TRACTS (N = 363)*					SUBURBAN TRACTS (N = 146)				
	1960	1970	Change 60-70			1960	1970	Change 60-70		
Selected Characteristics	$\overline{X}$	$\overline{X}$	$\overline{X}$	SD	R_{60-70}	$\overline{X}$	$\overline{X}$	$\overline{X}$	SD	R_{60-70}
Total Housing Units	1430	1444	+14	389	.93	1827	2277	+499	1010	.70
Housing Density	9315	8975	–339	2911	.93	2685	2975	+290	640	.96
Number of Single-Family Dwellings	324	332	+8	162	.95	1250	1387	+137	554	.87
Number of Apartments (3 or more units)	824	846	+22	372	.92	349	625	+276	522	.60
Number of Owner-Occupied Dwellings	456	485	+28	141	.97	1253	1525	+273	621	.84
Number of Renter-Occupied Dwellings	916	961	+45	343	.93	575	835	+259	565	.60
Total Population	4163	3987	–176	1225	.90	5922	6976	+1045	3174	.76
Population Density	26614	24449	–2165	11021	.83	8044	8388	+345	1287	.97
Median Family Income	6550	9813	+3263	2337	.69	9119	13957	+4838	1708	.96
Number Nonwhite	1059	1454	+395	1707	.73	158	300	+142	489	.83
Number Single	741	861	+120	344	.84	880	1296	+416	610	.53
Number Families With Children Under 18	564	385	–180	296	.75	864	892	+29	513	.76
Number Families Without Children Under 18	501	364	–139	271	.75	607	724	+118	320	.55

a. These figures are based on a weighted sample of central-city tracts. Low-income/high-density tracts and high-income/low-density tracts were under-sampled in the city. By weighting these tracts by a factor of 1.78, these statistics describe a representative sample of central-city tracted areas. The weighting procedure produces an artificially inflated sample size. The already representative suburban sample of tracts required no weighting.

TABLE 12.4 Effects of 1960 Zoning on Changes in Housing Density 1960-1970 for Central-City and Suburban Tracts (Dependent Variable is 1970 Housing Density/Units per Square Mile)[a]

Independent Variables	City	Suburb
1960 Housing Density (Housing Units per $Mile^2$)	0.87* (.03)	0.95* (.03)
Exclusionary Zoning Index[b]	–205.35 (125.68)	–253.01* (50.61)
% Zoned for Business	–52.18* (11.75)	–22.15* (3.93)
% Zoned for Manufacturing	–26.58* (7.84)	–4.18 (2.93)
Constant	3262.09* (727.79)	2325.74* (413.58)
R^2 =	.87	.94
N =	(237)	(144)

*Significant at .05 level.
a. Index ranged from 1 to 11, where 1 is permissive residential zoning and 11 is exclusive residential zoning.
b. Shown are regression coefficients with their associated standard errors in parentheses.

EFFECTS OF ZONING ON LOCAL HOUSING COMPOSITION

Although the results of Table 12.2 indicate that the zoning classifications of tracts were reflected in the housing densities and housing stock compositions of those tracts in 1960, the issue we are concerned with is how those zoning regulations affected change in the period from 1960 to 1970. Do tracts with high exclusionary ratings become more dense in 1970 than one would expect on the basis of their 1960 densities? Are there corresponding changes in the expected housing stock composition? These issues are addressed in Tables 12.4 and 12.5.

In Table 12.4 the effects of zoning on the 1970 densities of tracts are shown. Suburban tracts become less dense than expected, the higher the Exclusionary Zoning Index for that tract. Thus each additional level in average index values leads to about 253 fewer dwelling units per square mile in 1970.[20] The coefficient for Chicago tracts is about the same level, but does not reach statistical significance.

The coefficients for percentages of tract areas zoned for business and manufacturing also show that the larger the percentage of land zoned for those purposes in 1960, the less dense the housing than expected in 1970. The coefficients for business zoning are significant in both Chicago and the suburbs, but only in Chicago for manufacturing zoning. Each percentage of business zoning meant 52 fewer housing units per square mile in Chicago and 22 fewer units per square mile in the suburbs in 1970, possibly reflecting the tendency for business establishments to drive out residential uses over time. These coefficients may represent the classical ecological process of business and manufacturing competing successfully for land use but, within our framework, *especially when allowed to compete* by appropriately permissive zoning.

Table 12.5 disaggregates these general effects by estimating separately the impacts of zoning on 1960-1970 changes in the number of single-family dwellings, the number of multiple-family dwellings (three or more units per structure), the number of owner-occupied units, and the number of renter-occupied units.

Included in each model are measures of tract areas (square miles) and measures of the changes in the total number of dwelling units contained within each tract over the decade. Tract area was included because (a) the census's method of defining these units (by population size) tended to covary with tract zoning and (b) large tracts simply contain larger land areas available for potential development.[21] Including the change in the total number of housing units over the decade accounts for general housing market forces so that the zoning effects are net of the tendency for some tracts to have experienced more growth than others. The zoning coefficients thus represent the degree to which zoning influences the *type* of housing growth or the mix of housing within tract areas net of changes in the total housing stocks.

Within both the city and the suburbs, zoning directly influenced the type of housing development experienced from 1960 to 1970. More single-family dwellings tended to be added the more exclusively zoned a tract was, while more multifamily dwellings tended to be built within more permissively zoned tracts.

Note that these effects are quite substantial and are independent of the housing stock growth trends within tracts. Thus for each additional unit of the Exclusionary Zoning Index, a tract in Chicago experienced an additional 42 single-family homes and in the suburbs about 59 additional single-family homes. In other words, two tracts of the same area that were 10 units apart on the index and which grew at the same rate between 1960 and 1970 experienced in 1970, on the average, a difference of between 420 and 590 single-family homes. Correspondingly large dif-

TABLE 12.5 Effects of Zoning on Housing Changes 1960-1970 for Central-City and Suburban Tracts (Dependent Variables are Number of Housing Units 1970 for Selected Types of Housing)[a]

	Number of Single-Family Dwellings		Number of Multifamily Dwellings[b]		Number of Owner-Occupied Housing[c]		Number of Renter-Occupied Housing	
	City	Suburb	City	Suburb	City	Suburb	City	Suburb
1960 Housing Characteristic[d]	.78* (.04)	.92* (.03)	1.00 (.01)	1.06* (.04)	.92* (.03)	.93* (.03)	1.08* (.01)	1.10* (.04)
Exclusionary Zoning Index[e]	41.93* (6.48)	58.93* (14.48)	–21.86* (7.82)	–45.04* (14.55)	15.09* (6.01)	36.24* (14.04)	–16.37* (5.88)	–18.36 (15.57)
% Zoned for Business	.35 (.57)	1.83 (1.32)	.81 (.66)	–1.64 (1.26)	.02 (.54)	.006 (1.34)	–.43 (.50)	–.51 (1.26)
% Zoned for Manufacturing	–.90* (.39)	–.18* (.95)	–.37* (.43)	–.48 (.91)	–.54 (.40)	–.16 (.96)	.26 (.33)	–.51 (.90)
Change in Total Housing 1960-1970	.06* (.02)	.47* (.02)	.92* (.03)	.53* (.02)	.19* (.02)	.54* (.02)	.88* (.02)	.58* (.02)
Tract Area (square miles)	430.76* (60.18)	18.56* (4.78)	–177.95* (54.37)	–15.84* (4.59)	338.27* (56.50)	21.75* (4.87)	–226.60* (41.89)	–17.50* (4.55)
Constant	–160.17* (26.67)	–386.60* (91.54)	150.71* (39.03)	352.94* (108.33)	–42.34 (24.68)	138.90 (93.76)	73.25* (30.53)	95.20 (122.13)
R^2 =	.93	.97	.98	.92	.98	.98	.99	.94
N =	(237)	(144)	(237)	(144)	(237)	(144)	(237)	(144)

*Significant at .05 level.

a. Shown are regression coefficients with their associated standard errors in parentheses.

b. Number of units within structures containing three or more dwelling units.

c. Owner-occupied units includes condominium units.

d. The associated 1960 housing measures of each respective dependent variable.

e. Index ranges from 1 to 11, where 1 is permissive residential zoning and 11 is exclusive residential zoning.

ferences were also shown in multifamily dwellings, with the coefficients of 22 fewer such units in Chicago and 45 in the suburbs for each higher level of the Exclusionary Zoning Index.[22]

Since tenure and dwelling types are closely related, the two regression equations involving owner-occupied and renter-occupied units reflect largely the same trends as noted above, although at reduced levels.[23] Indeed, since the Chicago SMSA during that period experienced a growth in condominiums (owner-occupied units in multiple-unit structures), this lack of direct correspondence is partly to be expected on that basis alone.

Neither central-city nor suburban nonresidential zoning substantially influenced local housing development. The overall absence of direct effects of nonresidential zoning appear to reflect the lack of residential development within areas designated for commercial or industrial purposes.

It should also be noted that these four regressions all have R^2 values in excess of .90. Since most of the variance among tracts in 1970 is explainable on the basis of tract variance in 1960, such high values are to be expected. Yet given the correlations shown in Table 12.3, it is also clear that zoning does contribute significantly to the explanation of tract status in 1970.

EFFECTS OF ZONING ON LOCAL SOCIOECONOMIC COMPOSITION

By regulating neighborhood housing composition, zoning also indirectly regulates local socioeconomic composition. Obviously, lower-income families cannot choose to live within neighborhoods restricted to single-family housing if they cannot afford the consequent housing prices.[24] Because race and class are linked, restrictive zoning practices are also believed to be motivated by desires to racially segregate. Zoning has been viewed as an indirect legal method for ensuring that neighborhood racial segregation persists over time.

Table 12.6 models the effects of zoning on changes in tract median family income levels and changes in the number of nonwhite tract residents from 1960 to 1970. The change in the total population of tracts over the decade is included in the model of racial change in order to separate increases or decreases in nonwhites from overall population growth or decline.

Central-city zoning did not directly influence 1960-1970 changes in average household income levels of tract residents. Suburban zoning, however, clearly worked to increase neighborhood income segregation. Higher-income families tended to move more to exclusively zoned

TABLE 12.6 Effects of Zoning on Tract Socioeconomic Change 1960-1970 for Central-City and Suburban Tracts

	(Dependent Variable is Median Family Income 1970)[a]		(Dependent Variable is Number of Nonwhite 1970)[a]	
Independent Variables	**City**	**Suburb**	**City**	**Suburb**
1960 Socioeconomic Characteristic	1.16*	1.33*	.95*	1.18*
	(.09)	(.04)	(.04)	(.07)
Exclusionary Zoning Index[b]	39.19	248.07*	–67.30	–37.50
	(110.83)	(83.35)	(57.40)	(33.78)
% Zoned for Business	2.69	–7.17	4.09	–3.10
	(10.16)	(7.48)	(5.59)	(3.34)
% Zoned for Manufacturing	–2.97	–1.18	–3.02	–.90
	(6.73)	(5.63)	(3.41)	(2.52)
Change in Total Population 1960-1970[c]	–	–	.88*	.02
	–	–	(.07)	(.02)
Tract Area (square miles)	819.86	80.71*	1,547.46*	–4.93
	(812.31)	(23.82)	(437.47)	(12.14)
Constant	1,937.10*	177.12	417.68	379.57
	(709.11)	(539.57)	(277.98)	(235.20)
R^2 =	.51	.61	.72	.70
N =	(237)	(144)	(237)	(144)

*Significant at .05 level.
a. Shown are regression coefficients with their associated standard errors in parentheses.
b. Index ranges from 1 to 11, where 1 is permissive residential zoning and 11 is exclusive residential zoning.
c. Change in total population not included in model of tract income level changes.

residential areas while lower-income families tended to move more to less restrictively zoned areas.

Neither central-city nor suburban zoning worked to increase neighborhood racial segregation. The negative, though insignificant, coefficients associated with exclusionary zoning, however, are consistent with the interpretation that zoning did not interfere with maintaining established patterns of racial segregation.

The findings of Table 12.6 indicate that zoning apparently does not directly influence changes in the racial composition of residential areas. However, the *indirect* effects are likely to be considerable. Thus by influencing the density of dwellings and the types of structures that may be built in residential areas, thereby affecting the distribution of housing prices, zoning undoubtedly has considerable indirect effects. As we have seen, while zoning does not enhance racial segregation, neither does it lower segregation levels. The pattern of racial segregation shown so dramatically in Table 12.3 is left untouched by zoning regulations. Indeed, the influence of suburban zoning in increasing neighborhood economic segregation suggests that restrictive suburban zoning practices may have worked to maintain already established socioeconomic differences *between* the central city and the suburbs.

EFFECT OF ZONING ON LIFE-CYCLE COMPOSITION

Since housing requirements change as households go through typical life-cycle stages, zoning may affect life-cycle-related aspects of population composition by restricting or enhancing appropriate housing types. Thus a neighborhood that has small one-bedroom apartments is hardly regarded as a desirable location for couples with several school-age children. Similarly, zoning that permits only single-family homes on large lots may attract to a residential area couples with many school-age children.

Census tract statistics allow only indirect indicators of life-cycle stages. Table 12.7 contains three separate indicators of tract life-cycle composition: changes in the number of single adults, changes in the number of childless families, and changes in the number of families with children. The first equation includes changes in tract adult population to hold constant overall increases in adults. The remaining models include changes in total population. Again, these population change characteristics account for population trends *per se*, so that the effects of zoning are over and above general population trends.

TABLE 12.7 Effects of Zoning on Family Life-Cycle Compositional Changes Within Tracts 1960-1970 for Central-City and Suburban Tracts[a]

	(Dependent Variable is Number of Single Persons 1970)		(Dependent Variable is Number of Families Without Children 1970)		(Dependent Variable is Number of Families w/Children Under 18 1970)	
Independent Variables	**City**	**Suburb**	**City**	**Suburb**	**City**	**Suburb**
1960 Population Characteristic[b]	1.13*	1.11*	.56*	1.12*	.46*	1.21*
	(.03)	(.05)	(.03)	(.11)	(.03)	(.08)
Exclusionary Zoning Index[c]	−26.47*	7.00	15.40	56.85*	23.44*	19.90
	(8.28)	(11.75)	(8.66)	(19.89)	(7.38)	(25.79)
% Zoned for Business	−3.07*	−1.41	.18	1.09	−.36	2.62
	(.81)	(1.16)	(.88)	(1.94)	(.74)	(2.58)
% Zoned for Manufacturing	−1.04*	1.01	−1.44*	−1.06	−1.22*	−1.86
	(.53)	(.85)	(.57)	(1.38)	(.46)	(1.84)
Change in Total Population 1960-70[d]	.36*	.25*	.06*	.04*	.03*	.08*
	(.02)	(.01)	(.01)	(.01)	(.01)	(.01)
Tract Area (square miles)	−80.28	−2.36	239.03*	5.95	303.96*	26.98
	(68.37)	(3.97)	(68.48)	(6.91)	(58.07)	(9.44)
Constant	244.94*	42.49	10.28	−370.01*	14.94	−426.57*
	(43.93)	(98.00)	(46.24)	(175.65)	(37.60)	(187.33)
R^2 =	.92	.95	.68	.51	.69	.77
N =	(237)	(144)	(237)	(144)	(237)	(144)

*Significant at .05 level.
a. Shown are regression coefficients with their associated standard errors in parentheses.
b. The associated 1960 population measure of each respective dependent variable.
c. Index ranges from 1 to 11, where 1 is permissive residential zoning and 11 is exclusive residential zoning.
d. Examination of the change in the number of single persons accounts for the change in the total number of adults (14 years and older) from 1960 to 1970.

Central-city zoning worked to distribute urban residents according to stages in the family life cycle. The number of single persons tended to increase more in permissively zoned residential areas, while the number of married people with or without children tended to increase more in restrictively zoned neighborhoods.[25]

Suburban zoning's influence on change in tract family life-cycle composition was less pronounced. Tracts gained neither more single persons nor families with children due to local zoning. Rather, exclusively zoned suburban neighborhoods tended to have larger increases in numbers of childless households.

These city-suburb differences emphasize the difference between urban and suburban residential environments. Permissively zoned central-city areas accompanied by traffic, noise, and an absence of open space may have been unsuitable for raising young children; perhaps it was particularly salient for urban families with children to move to restrictively zoned single-family neighborhoods. By contrast, the more exclusively zoned suburbs were less dense and did not pose the externalities offered by central-city neighborhoods, so that suburban families were less restricted in their choice of a suitable environment for rearing children.[26] The attraction of childless suburban families to more exclusively zoned neighborhoods may reflect that these couples (professional working couples or older couples in the "empty nest" stage) simply had more money to spend on housing than did growing families.

CONCLUSIONS AND IMPLICATIONS

According to one writer on zoning, "We had better recognize that the humble zoning ordinance is probably of direct concern to more people than any other statute" (Sussna, 1966). This examination of zoning provides good reasons for such concerns. Zoning is an important local political method for shaping the housing and population composition of metropolitan neighborhoods. An overview of the findings presented elucidates the way in which this process works.

(1) By limiting housing density, zoning directly influences increases in the availability of rental housing.

(2) Zoning methods create increasingly segregated neighborhoods in terms of housing tenure. Restrictively zoned neighborhoods tend to become increasingly neighborhoods of homeowners, while less restrictively zoned neighborhoods become more intensively neighborhoods of renters.

(3) Central-city zoning does *not* tend to increase neighborhood socioeconomic segregation. Urban zoning simply increases trends in which households are sorted according to stages in the family life cycle.

(4) Suburban zoning tends to increase neighborhood income segregation. Although zoning does not directly influence neighborhood racial segregation, zoning may maintain already established patterns by including or excluding the types of people who can afford to live within differently regulated areas.

Such findings should not be surprising. In fact, these are the general goals of the zoning process. Nevertheless, it is important to recognize that the intent of this chapter is not to reify a planning method but, rather, to show how local decisions about land use in one major metropolitan region affect trends in particular social and spatial patterns. That these findings were "as expected" only shows how engrained contemporary values are concerning housing, neighborhoods, and the like and ensures that the basic principles underlying the zoning process remain unchallenged.

Zoning is a local political process. Planning decisions are made by representatives of the local community: by aldermen, appointed or volunteer planning board members, or representatives of the general community. The forcefulness of zoning decisions in shaping the metropolitan landscape is not due to the emergence of forces resulting from unconcerted movements of people or from the buying or selling of property. Zoning is a concerted effort on the part of communities or neighborhoods to plan local development. It is conscious, intentional, and has clearly intended results.

Of course, the process by which zoning decisions are made is not simply technocratic or democratic. That zoning decisions are political means that there are costs and benefits to these decisions. The outcome from struggles over zoning may be the result of power inequities. Developers may be more effective in obtaining zoning changes compared to other citizens. Low-income people within or outside local boundaries may be unable to do the same because of the lack of financial or political resources. That the way in which zoning decisions are made may be at best, unfair or, at worst, corrupt does not change the fact that zoning decisions ensure that neighborhoods develop in particular ways over and above market forces, neighborhood accessibility, and the like.

That urban (central-city) and suburban zoning similarly influences housing development does not mean that zoning works the same within city centers as within peripheral suburban communities. The more permissively zoned central city was the receiver of less housing development and more nonwhite, lower-income people than was suburbia. While urban zoning regulated the density and types of new development, it did not (except for families) work to increase local socioeconomic or racial segregation. Possibly within the already built-up urban environment, zoning may simply reflect existing land use. Hence, the way in which land use is regulated within the central city may ensure that neighborhoods remain essentially the same and continue to gain comparable densities and housing over time.

This is not true for suburban communities which gained large numbers of people and housing and continued to become more wealthy relative to the central city during the period examined. The less dense and less developed suburban areas were predominantly restrictively zoned, at least in terms of restricting any type of housing development to the single-family home. As in the city, such zoning effectively segregated residential land use. At the same time, these zoning restrictions operated as what is termed an "invisible wall" and worked to increase the segregation of families by income.

Since this study examined zoning in the city compared to zoning in the suburbs, the issue of suburban exclusion cannot be directly addressed. Given the different levels of zoning restrictiveness within these two areas and the differential effects of this process, suburbs clearly may be excluding potential newcomers from their boundaries. That suburban neighborhoods effectively excluded families by income because of local zoning supports the idea that suburban zoning not only distributed population within these peripheral areas but also excluded them from these areas entirely.

The point that suburban zoning may in fact be exclusionary, ensuring the separation of the rich from the poor, homeowners from renters, single-family homes from apartments means that zoning is a policy which directly influences the development of housing for different types of people. The inability to provide adequate housing for lower-income families, racial minorities, and others may stem directly from the zoning process.

Moreover, these effects of zoning show that housing and population are not simply distributed according to ecological forces or market mechanisms; and it is for this reason that the present social-spatial pattern of the metropolis is not a "given" determined by emergent forces but a pattern that can be changed. Similar types of people and housing are spatially segregated into different areas not simply because of ethnic affinities, competition over land, or neighborhood accessibility, but partly because of local decisions to use land in particular ways.

Urban ecology emphasizes the social or subsocial forces conditioned by the market as driving urban social-spatial change. This study illustrates the inadequacy of this framework for understanding how neighborhoods develop and why and where people live where they do. We look toward a new urban ecology which incorporates political processes and their consequences for the quality of urban life.

APPENDIX
Municipalities and Counties Containing Tracted Areas

Cities	Peripheral Municipalities	Counties
Chicago	Arlington Heights	Cook County
	Aurora	DuPage County
	Bedford Park	
	Berwyn	
	Burr Ridge	
	Calumet City	
	Chicago Heights	
	Cicero	
	Des Plaines	
	Downers Grove	
	Elgin	
	Elmhurst	
	Evanston	
	Evergreen Park	
	Forest Park	
	Glencoe	
	Glen Ellyn	
	Glenview	
	Highland Park	
	Hometown	
	Joliet	
	LaGrange Park	
	LaGrange	
	Lansing	
	Lyons	
	Maywood	
	Melrose Park	
	Morton Grove	
	Naperville	
	Niles	
	Oak Park	
	Park Ridge	
	River Forest	
	River Grove	
	Skokie	
	Steger	
	Stickney	
	Waukegan	
	Wilmette	

NOTES

1. McKenzie expresses a similar idea: "Human ecology is fundamentally interested in the effect of position in both time and space upon human institutions and human

behavior." He defines "position" as "the place relation of a given community to other communities, also the location of the individual or institution within the community itself" (in Park, 1967: 64).

2. For example, Schwirian's collection of writings on urban ecology (1974b) provides no serious discussions of the politics surrounding the urban planning process. Berry and Kasarda (1977) appear to dismiss planning as politics within the United States, arguing that in contrast to nationwide or statewide planning, localized planning decisions reflect competitive market processes. There are, however, some shifts in the direction of looking within the political context for mechanisms impeding or facilitating residential mobility with the goal of linking policy initiatives to improving the matching of needs and opportunities at the local level. This focus on the outcomes of residential mobility rather than the mobility process for the formulation of public policy can be found in Clark and Moore, *Residential Mobility and Public Policy* (1980).

3. According to geographer David Harvey, "Inherent in social processes lies the question of spatial organization. Externality effects are localized, so are job and housing opportunities, resource benefits, communication links, etc. Political power is partly areally based. *Many of the hidden mechanisms for redistributing income come to fruition in the act of location.*" (Harvey, 1973: 86).

4. Zoning is now so much a part of contemporary "good practice" in urban government that it is often required as a condition for participation in several federal community programs (for example, the requirement for a comprehensive plan as a prerequisite for federal urban renewal funds).

5. Police power pertains to the protection of the "general health, welfare and safety of the community." Theoretically, police power acknowledges the potential conflict between the free, unrestricted individual and the community. Under police power, individual freedom may be limited if a threat of harm exists for the community.

6. [The] segregation of residential, business, and industrial buildings will make it easier to provide fire apparatus suitable for the character and intensity of the development in each section; that it will increase the safety and security of home life; greatly tend to prevent street accidents, especially to children, by reducing the traffic and resulting confusion in residential sections; decrease noise and other conditions which produce or intensify nervous disorders; preserve a more favorable environment in which to raise children, etc. . . . [W]ith particular reference to apartment houses, it is pointed out that the development of detached house sections is greatly retarded by the coming of apartment houses, which has sometimes resulted in destroying the entire section for private house purposes; that in such sections very often the apartment house is a mere parasite, constructed in order to take advantage of open spaces and attractive surroundings created by the residential character of the district. Moreover, the coming of one apartment house is followed by others, interfering by their height and bulk with the free circulation of air and monopolizing the rays of the sun which otherwise would fall upon smaller homes, and bringing as their necessary accompaniments, the disturbing noises incident to increased traffic and business. . . . until, finally the residential character of the neighborhood and its desirability as a place of detached residences are utterly destroyed" (Euclid v. Ambler Realty Co., 1926).

7. Presumably, intervention on behalf of homeowners deprives no one of any significance (except would-be homeowners, poor families whose housing might be improved by lower price levels, and the like).

8. A negative externality or "nuisance" is a type of land use which is perceived to be harmful within an area. Economically, a negative externality is a harm which causes a decrease in the value of property (Mills, 1972).

9. According to the National Commission on Urban Problems, "there are sometimes important nonfiscal policies behind certain types of exclusionary land use decisions. 'Undesirable' people—minority groups and the poor—would not 'fit in' " (1968: 285).

10. This design has been successfully employed in studying the long-term (decade) effects of natural disasters (Wright et al., 1979).

11. Because tracts may change in population size, the Census Bureau redefines some tract boundaries at each census taking. These tracts were changed in two ways. First, a 1960 tract would be split into two, three, or more tracts in 1970. These split tracts could be added together in order to form a comparable area with their respective 1960 counterparts. Second, tract boundaries can be shifted, resulting in areas that are not comparable over time. While such boundary changes are usually minor, some are quite substantial. Therefore, we chose to include only perfect one-to-one tracts and perfect combinations of tracts and to drop tracts which did not form perfectly comparable areas.

12. Since Chicago tracts tended to be concentrated in more permissive zoning classifications, a straight random sample of tracts would have produced an uneven distribution of zoning classifications within thee city and hence restrict the variance in zoning among tracts. The sampling strategy employed tended to achieve a more even distribution of zoning and hence a larger variance in zoning practices than is representative of the city as a whole. In terms of the regression analyses used, this sampling strategy produces unbiased regression coefficients but lower sampling errors, hence enhancing the capabilities of rejecting Type II errors.

13. Unincorporated areas were governed by county zoning agencies.

14. Other analyses have demonstrated that specific regulations not accounted for in these methods for stratifying zoning categories tend to vary with the level of restrictiveness within each standard category. One may generally view any residential land use category as a package of corresponding regulations where both the type of land use permitted and the minimum lot area requirements set the limits on the absolute level of restrictiveness for each type of regulation (Shlay, 1981).

15. The seven standard nonresidential zoning districts are ranked according to potential negative externalities, as follows: Undeveloped Land, Office-Research, Limited Retail Trade, General Retail Trade, General Commercial, Light Manufacturing, and General Manufacturing.

16. The area measurement technique utilized was the "dot grid method." A dot grid is an overlay of uniformly distributed dots. To measure zoning districts we (1) outlined the tract boundaries onto each zoning map, (2) counted the number of dots within each zoning district within each tract, (3) found the total number of dots within the tract, and (4) formed a ratio of the number of dots in a particular zoning district to the total number of dots in the tract. For census tracts containing more than one local political jurisdiction, a slightly different method was used. Since each local zoning map was at a different scale, characterizing these tracts by their zoning was problematic because each map represented different amounts of land. For these tracts, the scales of each zoning map containing the parts of the tracted area were used to convert the number of dots to a common scale of measurement.

17. For each tract, the percentage zoned for each level of residential zoning restrictiveness (S-1, S-2, and so on) was weighted by its level of residential exclusion relative to other categories in the hierarchy (S-1=11, S-2=10, . . . M-4=1). These weighted percentages were combined into a composite score and divided by the total percentage of each tract zoned for residential land use.

$$\text{Exclusionary Zoning Index} = \frac{W_i P_i}{P_t}$$

where

W_i = weight assigned to land use category i, for example, Single Family–1 receives a score of 11

P_i = percentage of tract zoned for land use category i

P_t = total percentage of tracted area zoned for residential land use.

18. A score of zero is difficult to interpret within this scheme. Technically, zero indicates the most permissively zoned tracts. A score of zero, however, is obtained by tracts containing no land zoned for housing. While it can be argued that tracts containing only nonresidential zoning are, in fact, permissively zoned, the purpose of the exclusionary zoning index is to summarize the level of *residential* zoning exclusiveness within tracts. Therefore, for these purposes, tracts receiving a score of zero are included in this discussion. In later analyses, these tracts are not included in analyses in which the Exclusionary Zoning Index is used.

19. When other characteristics of tracts are entered into the regressions, these variables are exogenous to the 1960 characteristics.

20. Note that each unstandardized regression coefficient represents the average net increase (or decrease) in the number of housing units per square miles associated with a unit change in the corresponding independent variable. A "net decrease" does not mean that more exclusively zoned tracts experienced an absolute decrease in housing density, but that the increase in housing density was smaller in more exclusively zoned tracts compared to more permissively zoned tracts.

21. Since census tract boundaries are defined in terms of the number of people living within these areas, large tracts (in area) tend to contain fewer people per square mile and hence have lower housing densities than small tracts. (The correlation between tract area and housing density in 1960 was -.40). Including tract area in the model thus enables the estimation of zoning effects that are not artifacts of the way census tracts were defined.

22. Analyses on changes in tracts' total housing stocks show that more exclusively zoned tracts tended to gain more housing units than permissively zoned tracts, although these coefficients are not statistically different from zero. Consistent with zoning's impact on changes in tract housing density, tracts zoned heavily for business and manufacturing gained fewer housing units compared to residentially zoned areas. Zoning's influences on changes in tract total population tend to mirror zoning's impact on changes in the total number of housing units contained within these areas (Shlay, 1981).

23. Other analyses examine zoning's influence on changes in the cost of tract housing over the decade, specifically changes in median rent levels and median owner-assessed housing values. Rent levels increased more within exclusively zoned tracts, which limited the amount of land available for rental housing. Rent levels increased less within permissively zoned tracts with little or no restrictions on high-density development. Zoning's impact on housing value appreciation indicates that different processes were operating within Chicago and within the suburbs. Within suburban areas, housing values tended to appreciate more

within exclusively zoned residential tracts. Within Chicago, housing values tended to appreciate more within permissively zoned residential tracts. These different effects may indicate that within an already "built up" central city where vacant land is scarce, land values tend to increase at a greater rate within areas zoned to permit competition for high-intensity land use (Shlay, 1981).

24. Whether this constitutes class discrimination depends on whether housing prices are elevated above "free" market levels, an issue we cannot address with these data.

25. Whether these differential changes occur through population movements into or out of the tracts cannot be ascertained. Probably both processes are occurring at the same time: unattached adults moving into permissively zoned areas and families with children moving out.

26. The differences between central-city and suburban residential environments are suggested by the distribution of residential zoning within tract areas. Within the central city, tracts zoned for residential land use tended to be either entirely permissively zoned or entirely restrictively zoned. Within the suburbs, however, few tracts were entirely permissively zoned. Residentially zoned suburban tracts tended to be dominated by zoning for single-family dwellings and contained small amounts of land for high-density land use. Hence, suburban tracts with low Exclusionary Index Scores were likely to contain some single-family housing (Shlay, 1981).

REFERENCES

BABCOCK, R. F. (1973) Exclusionary Zoning: Land Use Regulation and Housing in the 1970s. New York: Praeger.

——— (1972) "The new Chicago zoning ordinance," in R. B. Andrews (ed.) Urban Land Use Policy. New York: Free Press.

——— (1969) The Zoning Game. Madison: University of Wisconsin Press.

BASSETT, E. M. (1936) Zoning: The Laws, Administration, and Court Decisions During the First Twenty Years. New York: Russell Sage.

BERGMAN, E. (1974) Eliminating Exclusionary Zoning: Reconciling Workplace and Residence in Suburban Areas. Cambridge, MA: Ballinger.

BERRY, B.J.L. and J. D. KASARDA (1977) Contemporary Urban Ecology. New York: Macmillan.

BEUSCHER, J. H., R. R. WRIGHT, and M. GITELMAN (1976) Cases and Materials on Land Use. St. Paul, MN: West Publishing.

BOHRNSTEDT, G. W. (1969) "Observations on the measurement of change," in E. F. Borgatta (ed.) Sociological Methodology. San Francisco: Jossey-Bass.

CASTELLS, M. (1976) "Theory and ideology in urban sociology," in C. G. Pickvance (ed.) Urban Sociology. New York: St. Martin's Press.

CLARK, W.A.V. and E. G. MOORE (1980) Residential Mobility and Public Policy. Beverly Hills, CA: Sage.

COLLIVER, A. and M. SEMYONOV (1979) "Suburban change and persistence." American Sociological Review 44: 480-486.

DANIELSON, M. B. (1976) The Politics of Exclusion. New York: Columbia University Press.

DAVIDOFF, P. and L. DAVIDOFF (1971) "Opening the suburbs: toward inclusionary land use controls." Syracuse Law Review 22: 509-535.

DELAFONS, J. (1969) Land Use Controls in the United States. Cambridge: MIT Press.

DOEBELE, W. A., Jr. (1972) "Key issues in land use controls," in R. B. Andrews (ed.) Urban Land Use Policy. New York: Free Press.

DOWNS, A. (1973) Opening Up the Suburbs. New Haven: Yale University Press.
——— (1970) Urban Problems and Prospects. Chicago: Markham.
EUCLID v. AMBLER REALTY CO. (1926) 272 U.S. 365
FITCH, R. (1977) "Planning New York," in R. E. Alcaly and D. Mermelstein (eds.) The Fiscal Crisis of American Cities. New York: Vintage.
FRIEDMAN, L. (1968) Government and Slum Housing. Chicago: Rand McNally.
GEORGE, R. D. (1971) "A survey of the judicial responses to exclusionary zoning." Syracuse Law Review 122: 537-633.
HAAR, C. (1959) Land Use Planning: A Case Book on the Use, Misuse, and Reuse of Urban Land. Boston: Little, Brown.
HARTMAN, C. W. (1975) Housing and Social Policy. Englewood Cliffs, NJ: Prentice-Hall.
HARVEY, D. (1973) Social Justice and the City. Baltimore: John Hopkins University Press.
HAWLEY, A. (1950) Human Ecology. New York: Ronald.
LEFEBVRE, H. (1977) "Reflections on the politics of space," in R. Peet (ed.) Radical Geography. Chicago: Maaroufa Press.
MANDELKER, D. R. (1972a) "A rationale for the zoning process," in M. Stewart (ed.) The City. Great Britain: Chaucer Press.
——— (1972b) "Control of competition as a proper purpose in zoning," in R. B. Andrews (ed.) Urban Land Policy. New York: Free Press.
MILLS, E. (1972) Urban Economics. Glenview, IL: Scott, Foresman.
MOLLENKOPF, J. (1978) "The postwar politics of urban development," in W. Tabb and L. Sawers (eds.) Marxism and the Metropolis: New Perspectives in Urban Political Economy. New York: Oxford University Press.
MOSKOWITZ, D. H. (1977) Exclusionary Zoning Litigation. Cambridge, MA: Ballinger.
National Commission on Urban Problems (1968) Building the American City. Washington, DC: Government Printing Office.
NATOLI, S. J. (1971) "Zoning and the development of urban land use patterns." Economic Geography 47: 171-184.
OLCOTT, George C. and Company (1970) Olcott's Land Values Blue Book of Chicago and Suburbs. Chicago: George C. Olcott and Company.
——— (1960) Olcott's Land Values Blue Book of Chicago and Suburbs. Chicago: George C. Olcott and Company.
PARK, R., E. W. BURGESS, and R. D. McKENZIE (1967) The City. Chicago: University of Chicago Press.
PENDLETON, B. F., R. D. WARREN, and H. C. CHANG (1980) "Correlated denominators in multiple regression and change analysis," in E. F. Borgatta and D. J. Jackson (eds.) Aggregate Data Analysis and Interpretation. Beverly Hills, CA: Sage.
PERIN, C. (1977) Everything in Its Place: Social Order and Land Use in America. Princeton, NJ: Princeton University Press.
ROSSI, P. H. (1980) Why Families Move. Beverly Hills, CA: Sage.
——— J. D. WRIGHT, S. R. WRIGHT, and E. WEBER-BURDIN (1978) "Are there long-term effects of natural disasters? estimating the effects of floods, hurricanes, and tornados occurring in 1960 to 1970 on U.S. counties and census tracts in 1970." Mass Emergencies 3: 117-132.
SCHAFER, R. (1975) "Exclusionary land use controls: conceptual and empirical problems in measuring the invisible wall." Discussion Paper D-75-2. Department of City and Regional Planning, Harvard University.

SCHWIRIAN, K. P. (1974a) "Some recent trends and methodological problems in urban ecological research," pp. 7-18 in K. P. Schwirian, Comparative Urban Structure. Lexington, MA: D. C. Heath.

——— (1974b) Comparative Urban Structure. Lexington, MA: D. C. Heath.

SCHULTZE, D. L., E. R. FRIED, A. M. RIVLIN, N. TEETERS, and R. REISCHAUER (1977) "Fiscal problems of cities," in R. E. Alcaly and D. Mermelstein (eds.) The Fiscal Crisis of American Cities. New York: Vintage.

SHLAY, A. B. (1981) "Zoning for whom: the impact of zoning on tract housing and population changes from 1960 to 1970 within the Chicago SMSA." Ph.D. dissertation, University of Massachusetts/Amherst.

SIEGAN, B. H. (1976) Land Use Without Zoning. Lexington, MA: D. C. Heath.

——— (1972) Other People's Property. Lexington, MA: D. C. Heath.

SMITH, M. (1979) "Critical Theory and urban political theory." Presented at the annual meeting of the American Political Science Association, Washington, D.C.

STERNLIEB, G. and J. HUGHES (1977) "Metropolitan decline and inter-regional job shifts," in R. E. Alcaly and D. Mermelstein (eds.) The Fiscal Crisis of American Cities. New York: Vintage.

SUDMAN, S. (1976) Applied Sampling. New York: Academic Press.

SUSSNA, S. (1970) Land Use Controls: More Effective Approaches. Washington, DC: The Urban Land Institute.

——— (1967) "Bulk control and zoning: the New York City experience." Land Economics 143: 158-171.

——— (1966) "Fifty years of zoning." American Bar Association Journal 152: 1030-1033.

——— (1961) "Zoning boards: in theory and in practice." Land Economics 37: 82-87.

TOLL, S. (1969) Zoned America. New York: Grossman.

WILLIAMS, N. (1975) American Land Planning Law. Chicago: Callaghan and Company.

——— (1966) The Structure of Urban Zoning and Its Dynamics in Urban Planning and Development. New York: Buttenheim.

——— and T. NORMAN (1971) "Exclusionary land use controls: the case of northeastern New Jersey." Syracuse Law Review 22: 475-507.

WRIGHT, J. D., P. H. ROSSI, S. R. WRIGHT, and E. WEBER-BURDIN (1979) After the Clean-Up: Long-Range Effects of Natural Disasters. Beverly Hills, CA: Sage.

Part VI

Epilogue

13

Federal Funding for Urban Policy Research

ROLAND J. LIEBERT

IN MANY RESPECTS social science research reflects its times. The choice of issues is a prime example. The abrupt political confrontation over issues of poverty, racial discrimination, and inequality, coincident with unprecedented rises in college enrollments and educational opportunity during the 1960s, no doubt propelled the research on stratification that continues with us today. The tightening worldwide interdependency that gave us OPEC and reopened China no doubt inspires the widening interest in historical world-systems theory. Urban riots, deterioration of the urban physical infrastructure, the accompanying rise in urban crime, mounting urban public programs, and the fiscal strains that followed no doubt fed the rediscovery of urban policy research.

It may even be said that broad research paradigms encompassing the methods and theory that guide research in the social sciences also reflect their times, although with delayed response. A recent example is the transformation out of the complacency of structural functionalism, with its abstract universalisms and satisfaction with soft data, into a more contentious pragmatism pitting ever harder data against a model of how things are said or meant to work. The latter type of research has best served the nonacademic social research sector and now is taking charge in academe as well. Here, as in much else in social science, the debt of basic research to applied research is great.

Another way social science research reflects its times is in its funding. Some may say that this drives all the other changes in what and how we do social science. While research will certainly be directed to where funds are plentiful, that leaves unanswered the question of why funds are more available for some types of work than for others, or why funds are available at all. A good part of the answer depends on what social science itself does and how that relates to the needs and interests of funders.

A case can be made that social science research has worked itself into several deep dilemmas with its funders. To spell out this case is not to chastise social science; indeed, the shortsightedness of funding sources is often more troublesome than the performance of social science. When the noble aim of pursuing truth is wedded to the noble aim of producing something of social significance, the results have power in a domain where power has many roots and voices.

Without going into details, it is clear that some social scientists helped pave the way for the social reforms of the long gone 1960s. Conservatives have neither forgotten nor forgiven the initial burst of academic acclaim for the federal programmatic efforts to reduce inequity. As for the liberals, they looked on as a new wave of hard-nosed researchers, schooled in the computer era, found time and again that the great social programs were rarely achieving their goals, often serving only to quiet their advocates, and costing far more than expected. So, the liberal voices were weakened and sent into disarray.

Oddly enough, it is precisely such a critical atmosphere that could best serve social science research at this time. A strengthened commitment to confront competing theories with the best data would only help social science. A strengthened commitment to address problems of social significance would, again, only help social science.

The Conference on Urban Policy Research forming the basis of this volume was predicated on the assumption that it could lay a foundation for one branch of the social sciences and some of the issues that it could address in the 1980s. The aim was to bring together a wide range of urban analysts who had competing theories about how and why urban social systems develop as they do, with special focus on the impact of public policy on such developments. From such a group, an assessment of common data needs was to be produced. With this assessment, the idea of developing a common data set was to be expanded into a proposal for large-scale funding.

For a number of reasons—not the least of which was the lack of funds in the foreseeable future—the full data assembly project was not funded. The notion of uniting urban researchers behind a common data collection effort was also resisted by some researchers. It was, perhaps, an idea before its time.

The time for social scientists to get a little more together, to find a way to address competing and common issues on some mutually acceptable standards or data, is rapidly drawing near. Diverse, highly original individual efforts will and must still be prized above all else, but it is folly to continue to pretend that urban social science—let alone all of social science—can thrive in the present atomized mode alone.

This problem has been met and, to some satisfactory degree, resolved in other fields of science. The astronomers have their five national laboratories with telescopes and another huge one on the drawing boards. These are available to all who have a study in mind. The physicists have their major experimental labs, and departments throughout the country cooperate on equipping and staffing major programs of research in chemistry and in the biological sciences.

Social scientists have little need for high-priced equipment, beyond the already available university computer facilities. But they do require huge data sets, for which the costs of production and data file preparation are so great as to require external funding.

Recognizing this fact, the National Science Foundation has, for nearly two decades, funded the General Social Survey by NORC to provide survey research data on the basic values and attitudes of Americans. This continuing effort has probably yielded more dissertations than any other social science data set, while providing some information for scores of professionally published papers. The more recent NSF commitment to the National Election Studies data series has led to a similar growth in research performance. The new project to produce a public use sample of individual cases from the 1940 and 1950 census returns is also expected to have a great impact on cumulative, interrelated lines of research. All of these projects are concerned with individuals, their attributes and attitudes. It is long past time that the social science research community put together a set of mutual data needs for social units, such as organizations or urban places, where individuals play out their lives.

To achieve such an end requires much more than the willingness of funders to support a major, long-range data collection and assembly effort. It requires as well a sense in the relevant research community that there is a genuine prospect for scientific progress as a result of having such a data set. It requires, to put it procedurally, a reasonably wide agreement on what is needed in such a data set. If this does not exist, it cannot be manufactured. The question the present volume addresses is whether such an agreement is forthcoming.

The participants took important steps toward identifying common data needs. In the absence of follow-up funding, however, the task of the conference remains unfinished. If this volume serves to continue the dialogue begun at the conference, its aims may yet be accomplished.

About the Contributors

WILLIAM T. BIELBY is working with James N. Baron on a study of economic segmentation and work organization, utilizing data on over 600 California firms. Their project involves reconstructing hierarchies of skill, promotion, and authority from data collected by the U.S. Employment Service from 1959 to 1980 in order to test competing accounts of economic segmentation.

STEPHEN C. BROOKS is a Ph.D. candidate in political science at Northwestern University, where he is completing a dissertation on the urban budgetary process. His interests also include analysis of the media and politics, organizational theory, and police politics. Before coming to Northwestern, he was director of off-campus programs at Coe College.

RONALD S. BURT is Assistant Professor in the Department of Sociology and Principal Investigator in the Survey Research Center at the University of California, Berkeley. His interest is the general development of structural sociology.

TERRY NICHOLS CLARK is Associate Professor of Sociology at the University of Chicago, and has taught at Columbia, Harvard, Yale, and the Sorbonne. He has worked at the Brookings Institution, the Urban Institute, U.S. Conference of Mayors, and Department of Housing and Urban Development, and has published almost 90 books and articles mainly concerned with urban decision-making and public policy.

LORNA CROWLEY FERGUSON is a Associate Study Director at the National Opinion Research Center affiliated with the University of Chicago. She is also a research associate with the National Council on Governmental Accounting. Her Ph.D. is in sociology from the University of Chicago. She is co-author of *Political Processes and Urban Fiscal Strain* soon to be published, and author of several articles on related topics.

ROGER FRIEDLAND is currently working on a corporate geography of the United States. His book, *Power and Crisis in the Central City,* will be published by Macmillan (London) in 1982.

LARRY E. HUCKINS currently has a joint appointment in the Center for the Study of Business and Government and the Department of Economics and Finance at Baruch College, City University of New York. His current research focuses on municipal government fiscal problems and factor demands, urban housing and tax abatement programs, and local government public choice problems.

JOHN D. KASARDA is Professor and Chair of the Department of Sociology at the University of North Carolina, Chapel Hill. He is co-author (with Brian J. L. Berry) of *Contemporary Urban Ecology* and author of numerous articles dealing with urban demographic-economic structure and urban public policy. His most recent research has focused on the implications of the spatial redistribution of people and jobs for national urban policies.

DAVID KNOKE is Professor of Sociology at Indiana University, specializing in political sociology. Under an NIMH research scientist development award he has been studying voluntary associations, resulting most recently in a monograph (with James R. Wood) *Organized for Action: Commitment in Voluntary Associations* (1981, Rutgers University Press). His current research, with Edward O. Laumann, is a comparative study of the social organization of national policy networks in the energy and health domains.

ROLAND J. LIEBERT is Associate Director of the Survey Research Laboratory and Associate Professor of Sociology at the University of Illinois, Champaign-Urbana. In August 1980 he completed a three-year term as Program Director for Sociology at the National Science Foundation. He is currently conducting a program of research on the political selectivity and resocialization of migrants in America, testing implications of public choice theory for regional political sorting. He is the author of *Disintegration and Political Action: The Changing Functions of City Governments in America* (Academic Press, 1976) and coeditor of *Power, Paradigms and Community Research* (Sage Publications, 1977).

ROBERT L. LINEBERRY is Dean of the College of Liberal Arts and Sciences at the University of Kansas in Lawrence. He was Professor of

Political Science at Northwestern University, where he and Herbert Jacob collaborated on a longitudinal study of crime in American cities.

ELINOR OSTROM is Chair of the Department of Political Science and Co-Director of the Workshop in Political Theory and Policy Analysis at Indiana University. Her research interests are in the study of urban service delivery and the effects of institutional arrangements on citizens, elected officials, bureau chiefs, and street-level bureaucrat behavior.

ROGER B. PARKS is Associate Director of the Workshop in Political Theory and Policy Analysis at Indiana University. Current research activities include the development and testing of organizational and behavioral models for public service delivery and applications of such models to the measurement of public agency performance.

PETER H. ROSSI is currently Professor of Sociology and Director of the Social and Demographic Research Institute at the University of Massachusetts at Amherst, and coeditor of *Social Science Research.* He has been on the faculties of Harvard University, Johns Hopkins University, and the University of Chicago, where he also served as Director of the National Opinion Research Center. He has been a consultant on research methods and evaluation to (among others) the National Science Foundation, the National Institute of Mental Health, the Federal Trade Commission, and the Russell Sage Foundation. His research has largely been concerned with the application of social research methods to social issues, and he is currently engaged in research on natural disasters and criminal justice. His most recent works include *Reforming Public Welfare* (with K. Lyall, 1974), *Prison Reform and State Elites* (with R. A. Berk, 1977), *After the Clean-Up* (with Wright et al., 1979), and *Money, Work and Crime* (with R. A. Berk and K. J. Lenihan, 1980). In 1979-1980 Professor Rossi was president of the American Sociological Association, and in 1981 he receive the Lazarsfeld Award for contributions to social research methods.

PAUL D. SCHUMAKER is Associate Professor and Chairman of the Department of Political Science at the University of Kansas. He has published a variety of articles in such journals as *Journal of Politics, American Journal of Political Science, Administration and Society,* and *Sociological Quarterly* concerning democratic theory and practices in American cities.

ANNE B. SHLAY is Assistant Professor in the Department of Consumer Economics and Housing at Cornell University. Her research includes the social and economic effects of land use regulations, tenant organizing, and the social psychology of housing and neighborhood preferences.

GEORGE S. TOLLEY is Professor of Economics and Director of the Center for Urban Studies at the University of Chicago. One of his recent books is *Urban Growth Policy in a Market Economy* (Academic Press, 1979).